Advances in Neuroethics

Series Editors

Veljko Dubljević
North Carolina State University
Raleigh, NC
USA

Fabrice Jotterand
Medical College of Wisconsin
Milwaukee
USA

Ralf J. Jox
Lausanne University Hospital and University of Lausanne
Lausanne
Switzerland

Eric Racine
IRCM, Université de Montréal, and McGill University
Montréal, QC
Canada

Advances in neuroscience research are bringing to the forefront major benefits and ethical challenges for medicine and society. The ethical concerns related to patients with mental health and neurological conditions, as well as emerging social and philosophical problems created by advances in neuroscience, neurology and neurotechnology are addressed by a specialized and interdisciplinary field called neuroethics.

As neuroscience rapidly evolves, there is a need to define how society ought to move forward with respect to an ever growing range of issues. The ethical, legal and social ramifications of neuroscience, neurotechnology and neurology for research, patient care, and public health are diverse and far-reaching — and are only beginning to be understood.

In this context, the book series "Advances in Neuroethics" addresses how advances in brain sciences can be attended to for the benefit of patients and society at large.

Members of the international editorial board:

More information about this series at http://www.springer.com/series/14360

Orsolya Friedrich • Andreas Wolkenstein
Christoph Bublitz • Ralf J. Jox • Eric Racine
Editors

Clinical Neurotechnology meets Artificial Intelligence

Philosophical, Ethical, Legal and Social Implications

 Springer

Editors
Orsolya Friedrich
Institute of Philosophy
FernUniversität in Hagen
Hagen
Germany

Christoph Bublitz
Faculty of Law
University of Hamburg
Hamburg
Germany

Eric Racine
University of Montreal and McGill
University
Institut de recherches cliniques de Montréal
Montréal, QC
Canada

Andreas Wolkenstein
Institute of Ethics, History and Theory of
Medicine
Ludwig-Maximilians-Universität
(LMU) München
Munich
Germany

Ralf J. Jox
Institute of Humanities in Medicine and
Clinical Ethics Unit
Lausanne University Hospital and
University of Lausanne
Lausanne
Switzerland

ISSN 2522-5677   ISSN 2522-5685   (electronic)
Advances in Neuroethics
ISBN 978-3-030-64592-2   ISBN 978-3-030-64590-8   (eBook)
https://doi.org/10.1007/978-3-030-64590-8

This Springer imprint is published by the registered company Springer Nature Switzerland AG
The registered company address is: Gewerbestrasse 11, 6330 Cham, Switzerland

Preface

This book has a somewhat longer history and many helping hands were needed to realize it. Therefore, we would like to thank all persons who contributed their share so that this volume could finally be published. Since most of the articles collected in this book originate from a conference entitled "Neurotechnology meets Artificial Intelligence. Ethical, social and legal implications of neurotech and AI" (held in Munich, May 8–10, 2019), we wish to thank all those who made the conference the refreshing, inspiring, thought-provoking, and enjoyable event that it was. The conference brought together a wide range of scholars with various disciplinary backgrounds (philosophy, law, social science, cognitive sciences, medicine) to discuss the multidimensional implications of neurotechnology and AI. It was mainly the outcome of Johannes Kögel's impressive organizing capabilities that the participants were able to experience a great conference, both academically and socially. The organizing team was supported by Nicola Williams, Natalie Kopczewski, and Armin Gruber who tirelessly helped in the background. Further, we would like to thank Georg Marckmann and the Institute of Ethics, History and Theory of Medicine at LMU Munich. Georg is head of the institute and continuously supported the conference and all activities around the INTERFACES project. Most importantly, we extend our gratitude to all speakers and authors without whose inspiring talks, smart contributions to the debates, and interest in the many facets of the conference's subject no such event would have been possible.

With regard to the realization of this book, we are deeply grateful to Meliz Kaygusuz and Bernadette Scherer. Due to their efforts we were able, among others, to overcome so many technical hurdles in the manuscript preparation and find all those tiny sources of potential errors that a book project usually hides. We are thankful for the proofreading services that Dorothea Wagner von Hoff has provided us with. Dorothea found many interesting, but clearly erroneous combinations of words that would have rendered some parts of the book extremely hard to read.

Finally, we would like to thank Sylvana Freyberg and the Springer team for their interest in publishing our book with them. They had to put a lot of patience in the project, so we are especially grateful for their continuous trust and interest. Moreover, the editors of the Springer series "Advances in Neuroethics"—Veljko Dubljevic, Fabrice Jotterand, Ralf J. Jox, and Eric Racine—accepted the inclusion of our book in the series, for which we are also very thankful.

Work on the book was funded by the Federal Ministry of Education and Research (BMBF) in Germany (INTERFACES, 01GP1622A) and by the Deutsche Forschungsgemeinschaft (DFG, German Research Foundation, 418201802), which we highly appreciate.

We now hope that readers find many important insights, points to consider, food for thought, and inspirations for their own work in the pages to come.

Hagen, Germany   Orsolya Friedrich
Munich, Germany   Andreas Wolkenstein

Contents

Contributors

Susanne Beck Institute for Criminal Law and Criminology, Leibniz University Hanover, Hanover, Germany

Christoph Bublitz Faculty of Law, University of Hamburg, Hamburg, Germany

Tom Buller, PhD Department of Philosophy, Illinois State University, Chicago, IL, USA

Sebastian Drosselmeier Munich Graduate School for Ethics in Practice, Ludwig-Maximilians-Universität (LMU) München, Munich, Germany

Tyr Fothergill De Montfort University, Leicester, UK

Orsolya Friedrich Institute of Philosophy, FernUniversität in Hagen, Hagen, Germany

Pim Haselager, PhD Donders Institute for Brain, Cognition, and Behaviour, Department of Artificial Intelligence, Radboud University, Nijmegen, The Netherlands

Marcello Ienca, PhD Department of Health Sciences and Technology (D-HEST), Swiss Federal Institute of Technology, ETH Zurich, Zurich, Switzerland

Ralf J. Jox Institute of Humanities in Medicine and Clinical Ethics Unit, Lausanne University Hospital and University of Lausanne, Lausanne, Switzerland

Johannes Kögel Institute of Ethics, History and Theory of Medicine, Ludwig-Maximilians-Universität (LMU) München, Munich, Germany

Hannah Maslen University of Oxford, Oxford, UK

Kevin McGillivray University of Oslo, Oslo, Norway

Giulio Mecacci, PhD Department of Artificial Intelligence, Radboud University, Nijmegen, The Netherlands

Department of Values, Technology and Innovation, Faculty of Technology, Policy and Management, TU Delft, Delft, The Netherlands

Eric Racine, PhD Pragmatic Health Ethics Research Unit, Institut de recherches cliniques de Montréal, Montreal, QC, Canada

Department of Neurology and Neurosurgery, McGill University, Montreal, QC, Canada

Department of Experimental Medicine (Biomedical Ethics Unit), McGill University, Montreal, QC, Canada

Department of Medicine and Department of Social and Preventive Medicine, Université de Montréal, Montreal, QC, Canada

Stephen Rainey, PhD Oxford Uehiro Centre for Practical Ethics, University of Oxford, Oxford, UK

Jean-Marc Rickli, PhD Geneva Centre for Security Policy (GCSP), Geneva, Switzerland

Matthew Sample, PhD Pragmatic Health Ethics Research Unit, Institut de recherches cliniques de Montréal, Montreal, QC, Canada

Department of Neurology and Neurosurgery, McGill University, Montreal, QC, Canada

Center for Ethics and Law in the Life Sciences, University of Hannover, Hannover, Germany

Jennifer R. Schmid Institute of Ethics, History and Theory of Medicine, Ludwig-Maximilians-Universität (LMU) München, Munich, Germany

Stephan Sellmaier Research Center for Neurophilosophy and Ethics of Neurosciences, Ludwig-Maximilians-Universität (LMU) München, Munich, Germany

Bernd Stahl De Montfort University, Leicester, UK

Georg Starke Institute for Biomedical Ethics, University of Basel, Basel, Switzerland

Mathias Vukelić, PhD Fraunhofer Institute for Industrial Engineering IAO, Stuttgart, Germany

Anna Frammartino Wilks, PhD Department of Philosophy, Acadia University, Wolfville, NS, Canada

Andreas Wolkenstein Institute of Ethics, History and Theory of Medicine, Ludwig-Maximilians-Universität (LMU) München, Munich, Germany

Introduction: Ethical Issues of Neurotechnologies and Artificial Intelligence

Orsolya Friedrich and Andreas Wolkenstein

Contents

Abstract

In this introduction to the volume, we present an overview of existing research on intelligent neurotechnologies, i.e., the combination of neurotechnologies with Artificial Intelligence (AI). Further, we present the ideas behind this volume and an overview of each chapter.

1.1 Neurotechnology + Artificial Intelligence = Intelligent Neurotechnologies (INT)

Imagine that the coffee machine in your kitchen starts brewing your urgently needed morning coffee as soon as you *think* the command "start the coffee machine" while you are still in bed. Is that realistic? Is it desirable? Using neurotechnologies, i.e.,

O. Friedrich (✉)
Institute of Philosophy, FernUniversität in Hagen, Hagen, Germany
e-mail: orsolya.friedrich@fernuni-hagen.de

A. Wolkenstein
Institute of Ethics, History and Theory of Medicine, Ludwig-Maximilians-Universität (LMU) München, Munich, Germany
e-mail: andreas.wolkenstein@med.uni-muenchen.de

technologies that lead to understanding, changing or interacting with the brain, combined with artificial intelligence (AI) might allow for such an application, even though many scientists doubt that technologies such as this one could be available in the near future. However, basic principles of brain-computer interfacing (BCI) have become reality and are currently the subject of intense research efforts [1–4]. BCIs measure brain activity and convert brain signals into computer commands, e.g., moving a cursor or a wheelchair [5, 6]. The most common way to measure brain activity is with non-invasive electroencephalography (EEG). BCIs use the power of thought or of focusing on a signal in order to give computational commands and require no neuromuscular innervation.

At the same time, BCIs and other neurotechnologies stand in relation with another emerging technology: AI. AI is already being used in many technologies to solve problems, which usually require human intelligence, such as reasoning, planning, and speech perception [7]. It is not a technology designed for a specific task, but cuts across all societal domains [8, 9] and comprises several technologies such as machine learning and artificial neural networks. The term "AI" thus denotes a variety of converging technologies that are used across many platforms and technologies. Kellmeyer [10] lists five different aspects: ubiquitous data collection, storage and processing of large amounts of data (big data), high performance analysis, machine learning, and interfaces for human-AI interaction.

AI is used in a number of ways in neuroscience and neurotechnology in the medical domain [11]. For example, computer vision capacities are being applied to detect tumors in magnetic resonance imaging (MRI) [12] or to detect anomalies in other kinds of data [13], e.g., EEG data [14–16]. These capacities lead to an improved diagnosis, prediction, and treatment of clinical pictures in a variety of medical domains [10]. In psychiatry, researchers have recently used AI to reach a biomarker-based diagnosis and determine therapy in patients with dementia, attention deficit hyperactivity disorder (ADHD), schizophrenia, autism, depression, and posttraumatic stress disorder (PTSD) [17–20]. AI that is used for speech recognition, in addition to many available data sources on the internet, helps researchers predict mental illness, for example [21].

Beyond its application in clinical research and therapy, AI is being used in combination with neurotechnologies. Big data and deep learning, for example, are promising trends that will influence the development of BCIs [22]. Among many other uses, these devices can be used by patients who suffer from amyotrophic lateral sclerosis (ALS) or severe paralysis in order to restore communication capacities and mobility, or in rehabilitation to facilitate the recovery process of patients after stroke [23–25]. With the help of AI, important BCI features such as signal processing and feature extraction can be improved [22]. Outside the strictly medical arena, EEG-based BCIs and other forms of AI-based neurotechnology are sold for entertainment purposes [26]. Facebook famously works with a typing-by-brain technology, which allows for a seamless social media experience [27]. Research behind this technology was already capable of showing how algorithms could decode speech in real time with a high amount of reliability [28]. Similarly, progress has been made in terms of facial recognition in EEG data [29]. BCIs, as well as other applications of (AI-enhanced) neurotechnology can also be found in military research. Warfighter enhancement is one motivation, but others include enhancing military equipment or deception detection [30–33].

In addition to technological development and progress, the number of articles, books, and events such as workshops or conferences that deal with the neuroethics of AI and neurotechnology is steadily increasing. Generally speaking, AI raises a host of original problems that can most aptly be summarized as "black box"-problems: It becomes increasingly difficult to supervise and control an AI's operation, because it manages its decision-making logic all by itself [34–37]. The combination of neurotechnologies and AI raises a host of further pressing problems. Yuste and colleagues [38] mention four broad areas of ethical concern: privacy and consent, agency and identity, augmentation, as well as bias. They propose various measures to address these issues, ranging from technological safeguards to legislation. For medical neurotechnology, a number of articles also emphasized problems regarding data protection and privacy as important issues to consider [39]. Moreover, questions of responsibility and shared agency are repeatedly brought up when it comes to neurotechnologies [40]. How BCIs affect agency and autonomy is another topic that drew attention to philosophers and ethicists [41, 42]. This body of research adds to more general approaches that examine the ethical quality of algorithms per se [9, 43]. Articles on issues such as hackability and problems derived from unwanted access to brain data [44] complement work that looks at specific forms of neurotechnology, e.g., in the medical, military, or consumer area [32, 33, 45, 46]. In addition, neurotechnology becomes increasingly interesting for political philosophers and others who approach INT with an eye on regulation questions and broader democratic worries [39, 47].

1.2 Novel Philosophical, Ethical, Legal, and Sociological Approaches to INT: An Overview

As this brief overview shows, many questions have already been addressed in the emerging literature both on technical issues and the normative implications of INT. Some of these questions have not been sufficiently or satisfyingly answered. Scholars from philosophy, sociology, and the law continue to exchange arguments and ideas while medical researchers, engineers, and computer scientists keep exploring new technologies and improve existing ones. The aim of this book is to provide a forum for the continuous exchange of these arguments and ideas. From a philosophical and ethical perspective, normatively relevant notions such as agency, autonomy, or responsibility have to be analyzed if humans interact with INT. This volume also asks, in a descriptive manner, how the reality of using INT would look like. It sheds light on the legal dimensions of INT. In addition, it explores a number of specific use cases, in that these concrete scenarios reveal more about the various domains of human agency in situations where technology and human-machine interaction play a distinctive role.

Accordingly, the methods used in this book vary considerably. They range from philosophical analysis, sociologically inspired descriptions, legal analysis, and socio-empirical research. This provides the book with the capacity to address a wide range of philosophical, normative, social, legal, and empirical dimensions of neuro-technology and AI. Most of the papers of this volume are the result of a conference that was held in Munich, in which the ethics of (clinical) neurotechnologies and AI were intensely discussed.[1]

The *first section* of the book reflects on some philosophically relevant phenom-ena and implications of neurotechnology use. From a philosophical and ethical per-spective, it must be asked how normatively relevant notions such as action, agency, autonomy, or responsibility can be conceptualized if humans act and interact with neurotechnologies. The most basic question is if BCI effects are actions at all and if there are normatively relevant differences between paradigmatic bodily actions and BCI-mediated actions. If there is no action or agency to be claimed, subsequent issues of autonomy and responsibility are affected, as well. Therefore, philosophical analyses of BCI use that focus on action-theoretical implications have emerged recently [41]. Two articles in this first section take this path.

Tom Buller analyzes the implications of BCI use for the nature of action. He claims that present BCI-mediated behavior fails to meet the necessary condition of intentional actions, namely the causation of an event and thus of bodily movement that is directly related to relevant beliefs and desires. Furthermore, he states that current BCI-mediated changes in the world do not qualify as non-deviant causal processes.

Sebastian Drosselmeier and *Stephan Sellmaier* also address the issue of action. However, they focus on the acquisition of a skill while using BCIs, which allows the user to make BCI-mediated changes in the world without performing a mental act. This would result—according to their argumentation—in the ability to perform BCI actions as basic actions. They also conclude that BCI users are able to differentiate between having a thought and an action relevant intention. Therefore, skilled users should be seen as competent and able to voluntarily control the BCI effects, which they cause in the world.

The concepts of action and agency are closely connected to the concept of auton-omy. Therefore, this suggests that some authors have recently also addressed the implications of BCI use on autonomy [42]. The first section of this volume also deals with this issue. Realizing the ability to act autonomously might be hampered or enhanced by using neurotechnologies.

Anna Wilks takes a closer look at the question of whether it would be a paradox or a possibility, following Kant, to augment autonomy through neurotechnologies. The paradox seems obvious at first hand: someone claims to augment autonomy with BCI use, but is able to perform self-legislation, whereas autonomous agency in a Kantian understanding requires that the person is not affected by external factors. Wilks, however, suggests that operating with a broader Kantian framework would

---

[1] https://neurotechmeetsai.wordpress.com/

allow integrating external components of BCIs into the understanding of self-legislation and thus avoid the paradox.

Pim Haselager, Giulio Mecacci, and *Andreas Wolkenstein* argue that BCIs, especially passive BCIs, shed new light on the traditional question of agency in philosophy. More precisely, they argue that the notion of ownership of action ("was that me?") might be affected by closely examining the action-theoretical implications of passive BCIs. If BCIs register intentions without the user being aware of this, and if they consequently act on them, then subconscious brain states may influence one's actions in a technology-mediated way. This observation serves as the basis for their plea to use passive BCIs, or what they call symbiotic technology, in experimentally guided thought experiments aimed at the study of the notion of agency. The authors suspect that by doing so, symbiotic technology may give new answers to how we must understand ownership of action and what consequences we have to expect.

Andreas Wolkenstein and *Orsolya Friedrich* contribute to the first section of the volume by summarizing the philosophical and ethical analysis that they described in their BCI-use analyzing project (Interfaces) and suggest some future directions for research and regulation of BCI development and use. They show that relevant results have been produced in recent philosophical, ethical, social, and legal reflections of BCI use. However, concluding results that could profoundly advise technology-regulating institutions or engineers are not present yet. Nevertheless, the development of AI-driven neurotechnologies are emerging and therefore, some preliminary ethically based regulatory framework is necessary. They suggest using procedural criteria as a first step.

Neurotechnology and AI also have broad social implications. These social implications not only include societal issues in general; certain areas of society, like research and medicine, are affected in a specific way. The *second section* of this volume focuses on some social implications of neurotechnology and AI use.

Matthew Sample and *Eric Racine* recall in their article that other emerging technologies, e.g., genomics or nanotechnology, have been promoted in ethics research in the past similar to the way that neural technologies are now. They address the question of how ethics researchers should deal with such research developments and question the significance of digital society for ethics research. They show how the significance of artificial intelligence and neural technologies, as examples of digital technologies, is affected by both sociological and ethical factors. They conclude that ethics researchers have to be careful in attributing significance and to reflect their own function in the process of attribution.

Johannes Kögel also focuses on BCI use from a sociological perspective. He shows that the BCI laboratory is not only a place to train this novel technology, but also a place of crisis management. The aim to discuss BCI use also as crisis management is to understand this social process and to increase sensitivity for the user experience. He argues that users currently experience BCI training and tasks as tedious and exhausting, because they have to make many "back-to-back decisions" for a long period of time and under immense time pressure, which is not common to activities in everyday life. His focus emphasizes the importance of developing BCI applications that allow for a more routine way of acting.

Jennifer R. Schmid and *Ralf J. Jox* further highlight the relevance and implications of the training process for the user experience in BCIs. They report on a qualitative interview study with healthy BCI users, e.g., neuro-gamers or pilots. The interviews show that the success of BCI use strongly depends on the motivation as well as the duration of training and that the time-consuming procedure of use results in discomfort and cognitive exhaustion.

This *second section* of this volume also approaches intelligent neurotechnologies from a legal perspective. The legal system faces the need to update some of its notions and regulatory action is needed to cover these new, neurotechnology-based forms of acting and acting together. BCIs also raise the question about mental privacy as well as data and consent issues.

Susanne Beck focuses on criminal law issues that result from neurotechnology use. She shows how neurotechnologies might lead to diffusion on the end of the victim, as well as the offender. Such diffusion would be important for criminal law, in that in traditional criminal law the roles of offender and victim are very clear. Therefore, criminalizing might lose some of its legitimacy. Another problematic diffusion in criminal law might occur, if there are no clear borders between the body and the mind.

Stephen Rainey et al. address further legally relevant issues, namely those related to data and consent in neural recording devices. They discuss whether current data protection regulation is adequate. They conclude that brain-reading devices present difficult consent issues for consumers and developers of the technology. They are also a potential challenge for current European data protection standards. Their use might become legally problematic, if the nature of the device results in an inability for the user to exercise their rights.

Finally, in the *third section* the book takes a closer look at neurotechnologies in their contexts of use. This section covers both the introduction of using neurotechnologies in various domains and an explication and discussion of their deeper philosophical, ethical, and social implications.

Ralf J. Jox discusses the ethical implications of the use of neurotechnologies and AI in the domain of medicine. He shows that such technology use challenges not only the patient–physician relationship, but also the whole character of medicine. He further highlights the potential threats to human nature, human identity, and the fundamental distinction between human beings and technological artifacts that could arise when AI technology with certain features is closely connected with the human brain.

The next contribution highlights one of these close connections of AI-neurotechnology and the human brain. *Stephen Rainey* discusses neuro-controlled speech neuroprosthesis from an ethical perspective. A speech neuroprosthesis picks out linguistically relevant neural signals in order to synthesize and realize, artificially, the overt speech sounds that the signals represent. The most important question in this special neurotechnology application is whether the synthesized speech represents the user's speech intentions and to what extent he can control the speech neuroprosthesis.

Georg Starke's contribution addresses another field of clinical neuroscience, namely the application of ML to neuroimaging data and the potential challenges of this application with regard to transparency and trust. He shows why transparency and trustworthiness are not necessarily linked and why transparency alone won't solve all the challenges of clinical ML applications.

Another field of application of neurotechnology and AI is their use in the military. *Jean-Marc Rickli* and *Marcello Ienca* discuss the security and military implications of neurotechnology and AI with regard to five security-relevant issues, namely data bias, accountability, manipulation, social control, weaponization, and democratization of access. They show that neurotechnology and AI both raise security concerns and share some characteristics: they proliferate outside supervised research settings, they are used for military aims, and they have a transformative and disruptive character. They highlight that it is extremely difficult to control the use and misuse of these technologies and call for global governance responses that are able to deal with the special characteristics of these technologies.

Finally, *Mathias Vukelić* directs our attention to a new research agenda for designing technology. Given the increasingly symbiotic nature of neurotechnology, where humans and technology closely interact, he emphasizes the need for a human-centered approach that puts human needs at the core. He attests that the detection of brain states, such as emotional or affective reactions, are of great potential for the development of symbiotic, interactive machines. Beyond assistive technology, this research leads to neuroadaptive technologies that are usable in a broad variety of domains. Vukelić argues that the primary goal of such an undertaking is the alignment of increasingly intelligent technology with human needs and abilities. While this could itself be viewed as following an ethical imperative, the author also stresses the wider ethical and societal implications of such a research agenda.

This short overview of existing research on intelligent neurotechnologies and of the articles in this volume offers a first insight into the emerging philosophical, ethical, legal, and social difficulties that we will have to face in the future and which require further conceptual as well as empirical research.

Acknowledgments Work on this paper was funded by the Federal Ministry of Education and Research (BMBF) in Germany (INTERFACES, 01GP1622A) and by the Deutsche Forschungsgemeinschaft (DFG, German Research Foundation)—418201802. We would like to thank Dorothea Wagner von Hoff for *proof reading* the article, Meliz-Sema Kaygusuz and Bernadette Scherer for formatting.

References

1. McFarland DJ, Wolpaw JR. EEG-based brain-computer interfaces. Curr Opin Biomed Eng. 2017;4:194–200.
2. Clerc M, Bougrain L, Lotte F, editors. Brain-computer interfaces 1. Foundations and methods. London: Wiley; 2016.
3. Clerc M, Bougrain L, Lotte F, editors. Brain-computer interfaces 2. Technology and applications. London: Wiley; 2016.

4. Graimann B, Allison B, Pfurtscheller G, editors. Brain-computer interfaces. Revolutionizing human-computer interaction. Berlin: Springer; 2010.
5. Shih JJ, Krusienski DJ, Wolpaw JR. Brain-computer interfaces in medicine. Mayo Clin Proc. 2012;87(3):268–79.
6. Graimann B, Allison B, Pfurtscheller G. Brain-computer interfaces: a gentle introduction. In: Graimann B, Allison B, Pfurtscheller G, editors. Brain-computer interfaces: revolutionizing human-computer interaction. Berlin: Springer; 2010. p. 1–27.
7. Luxton DD, editor. Artificial intelligence in behavioral and mental health care. Amsterdam: Elsevier; 2016.
8. Mainzer K. Künstliche Intelligenz. Wann übernehmen die Maschinen? Berlin: Springer; 2016.
9. Mittelstadt BD, Allo P, Taddeo M, Wachter S, Floridi L. The ethics of algorithms: mapping the debate. Big Data Soc. 2016;3(2):1–21.
10. Kellmeyer P. Artificial intelligence in basic and clinical neuroscience: opportunities and ethical challenges. e-Neuroforum. 2019;25(4):241–50.
11. Gunes O, Gunes G, Seyitoglu DC. The use of artificial intelligence in different medical branches: an overview of the literature. Med Sci. 2019;8(3):770–3.
12. Pereira S, Pinto A, Alves V, Silva CA. Brain tumor segmentation using convolutional neural networks in MRI images. IEEE Trans Med Imaging. 2016;35(5):1240–51.
13. Litjens G, Sánchez CI, Timofeeva N, Hermsen M, Nagtegaal I, Kovacs I, et al. Deep learning as a tool for increased accuracy and efficiency of histopathological diagnosis. Sci Rep. 2016;6(1):26286.
14. Tabar YR, Halici U. A novel deep learning approach for classification of EEG motor imagery signals. J Neural Eng. 2016;14(1):016003.
15. Schirrmeister RT, Springenberg JT, Fiederer LDJ, Glasstetter M, Eggensperger K, Tangermann M, et al. Deep learning with convolutional neural networks for EEG decoding and visualization. Hum Brain Mapp. 2017;38(11):5391–420.
16. Schirrmeister RT, Gemein L, Eggensberger K, Hutter F, Ball T. P64. Deep learning for EEG diagnostics. Clin Neurophysiol. 2018;129(8):e94.
17. Fakhoury M. Artificial intelligence in psychiatry. In: Kim Y-K, editor. Frontiers in psychiatry: artificial intelligence, precision medicine, and other paradigm shifts. Singapore: Springer Singapore; 2019. p. 119–25.
18. Vieira S, Pinaya WHL, Mechelli A. Using deep learning to investigate the neuroimaging correlates of psychiatric and neurological disorders: methods and applications. Neurosci Biobehav Rev. 2017;74:58–75.
19. Meyer-Lindenberg A. Künstliche Intelligenz in der Psychiatrie – ein Überblick. Nervenarzt. 2018;89(8):861–8.
20. Topol EJ. High-performance medicine: the convergence of human and artificial intelligence. Nat Med. 2019;25(1):44–56.
21. Eichstaedt JC, Smith RJ, Merchant RM, Ungar LH, Crutchley P, Preoţiuc-Pietro D, et al. Facebook language predicts depression in medical records. Proc Natl Acad Sci. 2018;115(44):11203.
22. Lin C, Liu Y, Wu S, Cao Z, Wang Y, Huang C, et al. EEG-based brain-computer interfaces: a novel neurotechnology and computational intelligence method. IEEE Syst Man Cybernet Mag. 2017;3(4):16–26.
23. Chaudhary U, Birbaumer N, Ramos-Murguialday A. Brain–computer interfaces for communication and rehabilitation. Nat Rev Neurol. 2016;12(9):513–25.
24. Salisbury DB, Parsons TD, Monden KR, Trost Z, Driver SJ. Brain–computer interface for individuals after spinal cord injury. Rehabil Psychol. 2016;61(4):435–41.
25. McFarland DJ, Daly J, Boulay C, Parvaz MA. Therapeutic applications of BCI technologies. Brain Comput Interfaces. 2017;4(1–2):37–52.
26. Blankertz B, Acqualanga L, Dähne S, Haufe S, Schultze-Kraft M, Sturm I, et al. The Berlin brain-computer interface: progress beyond communication and control. Front Neurosci. 2016;10:1–24.

27. Robertson A. Facebook just published an update on its futuristic brain-typing project. The Verge 2019 July 30.
28. Moses DA, Leonard MK, Makin JG, Chang EF. Real-time decoding of question-and-answer speech dialogue using human cortical activity. Nat Commun. 2019;10(1):3096.
29. Nemrodov D, Niemeier M, Patel A, Nestor A. The neural dynamics of facial identity processing: insights from EEG-based pattern analysis and image reconstruction. eNeuro. 2018;5(1):ENEURO.0358-17.2018.
30. Evans NG, Moreno JD. Neuroethics and policy at the National Security Interface: a test case for neuroethics theory and methodology. In: Racine E, Aspler J, editors. Debates about neuroethics. Cham: Springer; 2017. p. 141–57.
31. Tennison MN, Moreno JD. Neuroscience, ethics, and national security: the state of the art. PLoS Biol. 2012;10(3):e1001289.
32. Kotchetkov IS, Hwang BY, Appelboom G, Kellner CP, Connolly ES. Brain-computer interfaces: military, neurosurgical, and ethical perspective. Neurosurg Focus. 2010;28(5):E25.
33. Munyon CN. Neuroethics of non-primary brain computer interface: focus on potential military applications. Front Neurosci. 2018;12:696.
34. de Laat PB. Algorithmic decision-making based on machine learning from big data: can transparency restore accountability? Philos Technol. 2018;31(4):525–41.
35. Wachter S, Mittelstadt B, Floridi L. Transparent, explainable, and accountable AI for robotics. Sci Robot. 2017;2:eaan6080.
36. Pasquale F. The black box society. The secret algorithms that control money and information. Cambridge: Harvard University Press; 2015.
37. O'Neil C. Weapons of math destruction. How big data increases inequality and threatens democracy. New York: Crown; 2016.
38. Yuste R, Goering S, Agüera y Arcas B, Bi G, Carmena JM, Carter A, et al. Four ethical priorities for neurotechnologies and AI. Nature. 2017;551(7679):159–63.
39. Kellmeyer P. Big brain data: on the responsible use of brain data from clinical and consumer-directed neurotechnological devices. Neuroethics. 2018. https://doi.org/10.1007/s12152-018-9371-x.
40. Bublitz C, Wolkenstein A, Jox RJ, Friedrich O. Legal liabilities of BCI-users: responsibility gaps at the intersection of mind and machine? Int J Law Psychiatry. 2019;65:101399.
41. Steinert S, Bublitz C, Jox R, Friedrich O. Doing things with thoughts: brain-computer interfaces and disembodied agency. Philos Technol. 2019;32(3):457–82.
42. Friedrich O, Racine E, Steinert S, Pömsl J, Jox RJ. An analysis of the impact of brain-computer interfaces on autonomy. Neuroethics. 2018. https://doi.org/10.1007/s12152-018-9364-9.
43. Wolkenstein A, Jox RJ, Friedrich O. Brain–computer interfaces: lessons to be learned from the ethics of algorithms. Camb Q Healthc Ethics. 2018;27(4):635–46.
44. Ienca M, Haselager P. Hacking the brain: brain-computer interfacing technology and the ethics of neurosecurity. Ethics Inf Technol. 2016;18(2):117–29.
45. Ienca M, Haselager P, Emanuel EJ. Brain leaks and consumer neurotechnology. Nat Biotechnol. 2018;36:805.
46. Fiske A, Henningsen P, Buyx A. Your robot therapist will see you now: ethical implications of embodied artificial intelligence in psychiatry, psychology, and psychotherapy. J Med Internet Res. 2019;21(5):e13216.
47. Wolkenstein A, Friedrich O. Brain-computer interfaces: current and future investigations in the philosophy and politics of neurotechnology. In: Friedrich O, Wolkenstein A, Bublitz C, Jox RJ, Racine E, editors. Clinical Neurotechnology meets Artificial Intelligence: Philosophical, Ethical, Legal and Social Implications. Heidelberg: Springer; 2021.

Actions, Agents, and Interfaces

2

Tom Buller

Contents

Abstract

Ideally, a brain-computer interface (BCI) would enable bodily movement that is functionally and phenomenologically similar to "ordinary" behavior. One important element of this desired functionality is that the user would be able to control movement through the same types of mental activity that are used in "ordinary" behavior. For example, arm movement is caused by neural activity that underlies the conscious intention to move the arm. At present, however, the BCI-user has to learn to control movement by consciously imagining the movement, or by controlling neural activity that is only indirectly related to the intended movement. According to the standard account of action, a bodily movement qualifies as an action if its proximate cause is the conscious or unconscious intention to perform that movement. Since it can be argued that this condition is not met in the case of BCI-mediated behavior, an important question to ask is whether this type of behavior qualifies as intentional action.

T. Buller (✉)
Department of Philosophy, Illinois State University, Chicago, IL, USA
e-mail: tgbulle@ilstu.edu

© Springer Nature Switzerland AG 2021   11
O. Friedrich et al. (eds.), *Clinical Neurotechnology meets Artificial Intelligence*,
Advances in Neuroethics, https://doi.org/10.1007/978-3-030-64590-8_2

2.1   Introduction

A brain-computer interface (BCI) is a neuroprosthetic device that enables the control of bodily movement or an external device through the detection and decoding of neural activity. As the following case illustrates, significant progress in the development of BCI technology over the past years has helped increase the physical autonomy of individuals who have suffered a loss of motor function.

> [BK] has had electrical implants in the motor cortex of his brain and sensors inserted in his forearm, which allow the muscles of his arm and hand to be stimulated in response to signals from his brain, decoded by computer. After eight years, he is able to drink and feed himself without assistance. [1]

BCIs have been described as devices that translate thought into action [2–5]. This description seems appropriate since the movement of BK's arm and hand, for example, is neither a reflex nor did it just happen to occur; rather, the BCI detected and decoded BK's movement intentions and thereby effected the intended bodily movement. Accordingly, we might view BCIs as functional replacements for the damaged parts of the motor system, as novel realizers of the agent's movement intentions. In this regard BCIs present us with the latest—and most advanced—instance of replacement technology.

According to an influential and widely held view, physical actions are intentionally caused bodily movements. More precisely, the Causal Theory of Action (CTA) can be stated in the following way.

> (CTA) Any behavioral event A of an agent S is an action if and only if S's A-ing is caused in the right way and causally explained by some appropriately nonactional mental item(s) that mediate or constitute S's reasons for A-ing. ([6], p. 1)

The movement of BK's arm and hand counts as an action, therefore, because he wants to take a drink from his cup and the desire (and the attendant belief) causes the bodily movement. In this regard, actions are distinguished from "mere happenings"—bodily movements that lack this specific etiology. To say that a person's physical behavior is intentional is to say that it is causally related to their beliefs and desires. Tripping and falling over does not, therefore, count as an action since we can assume that the person did not have the belief and desire to trip and fall.

The matter is complicated, however, by the fact that not just any causal connection between intention and bodily movement will do. For we can imagine cases in which we would be reluctant to conclude that the person has acted even though bodily movement is causally related to the person's intentions.

> Bob desires and intends to shoot the sheriff, but this makes him nervous and causes his finger to cramp, which in turn causes the trigger to be pulled, resulting in the gun being fired and the sheriff being shot. ([7], p. 12)

Since the trigger being pulled was caused by Bob's nervousness, and his nervousness was caused by his intentional states, his bodily movement was causally related to his intentions. However, if we suppose quite plausibly that his nervousness was not itself intentional, then we can doubt that Bob intentionally shot the sheriff. To put the point in more theoretical terms: although the bodily movement matches Bob's original intention, it is not a *function* of his intention. As a consequence, we cannot exhaustively explain the trigger-pulling in terms of his beliefs and desires.

The above suggests that the causal process in physical action is of the right type if the intended bodily movement is a function of the person's beliefs and desires to perform that movement. If we adopt a broadly physicalist framework, then this is to say that an arm-raising, for example, qualifies as a physical action if it is brought about by the neurophysical state(s) that realizes the person's intention to move their arm. Unfortunately, this revised framework does not solve all our problems. For we can imagine cases in which our intuition is that the person has acted even though movement is not brought about by the appropriate neurophysical states.

> After suffering a severe spinal cord injury LC has lost a substantial degree of motor function. By concentrating on directional symbols displayed on a computer screen, LC is able to control the movement of a robotic limb with the aid of a BCI.

If we assume for the sake of argument that the neural activity underlying conscious attention is distinct from the neural activity underlying movement intention, then the causal process in this case is not of the right type. Nevertheless, LC would appear to be performing a physical action.

LC's case raises a number of important issues regarding the nature of physical action. First, we might ask whether, and under what conditions, the robotic limb counts as part of the body. Presumably, our answer to this question will depend in considerable part on the degree of functional and phenomenological similarity between control of the robotic limb and of "ordinary" arm movement—the greater the similarity, the greater the reason to conclude that that LC's robotic limb is "incorporated." If we conclude that robotic limb is not part of the body, then we can ask whether in moving the limb LC has performed a mental, rather than a physical, action or whether the movement is merely the effect of action.

Second, it can be claimed that the proximate cause of movement in LC's case is not *intention* to move the limb but *concentration* on a specific symbol. It is true, of course, that this event is not a mere happening since it is part of an intended causal process (unlike Bob's nervousness), but importantly the movement is not directly brought about by the neurophysical state(s) that realizes LC's intention to move the arm. Third, if it is the case that control of the robotic limb is substantially dissimilar to ordinary behavior, then it is not clear what is the nature and content of LC's beliefs and intentions which bring about movement. Ordinarily speaking, to intend to move my arm I must believe that my intention will bring the movement about, and that this movement is an arm movement. If the connection between intention and movement is unreliable, or the resultant movement is not of the right kind, then it is difficult to say exactly what I am intending.

2.2   BCIs and the Decoding of Movement Intention

An injury or disease that has damaged the spinal cord and has caused the loss of motor function may leave higher brain functions substantially intact. Motor function can be restored through an interface that bypasses the injury and connects the intact motor centers to an external device, robotic limb, or even the person's own body as in BK's case. A BCI designed to restore motor function decodes neural signals to extract voluntary motor commands that reflect the person's movement intentions, and then uses the process signal to control the external device or limb (robotic or "natural"). Typically, a BCI is composed of three components: a sensor to detect neural signals, a signal processor that converts neural activity into a command related to the desired action, and a device to effect action [10]. The BCI is able to detect motor commands from neural signals due to established correlations between neuronal firing rates and motor parameters like arm position, velocity, and joint torque [8]. Neuronal recording of motor commands has focused primarily on the primary motor cortex, although higher-level movement intentions and imagery, for example, imagined goals, trajectory and types of movement can also be decoded from the posterior parietal cortex [2, 8, 9].

BCIs can be categorized in a number of different ways [10]. First, we can distinguish the devices in terms of their level of invasiveness—whether they are placed on top of the scalp (EEG), subdurally on top of the cortex (ECoG), or inserted into the brain. Second, BCIs can be differentiated in terms of the type of signal recorded: field potentials (the summed electrical current from multiple neurons) from multiple recording sites (EEG and ECoG), or action potentials ("spikes") from single neurons or small groups of neurons. Although the less invasive devices can detect brain activity that correlates with visual stimuli and voluntary intention, for example, more specific and accurate details of action are obtainable from spiking activity. For instance, hand velocity and position, and movement goals can be detected from single cells in the motor cortex [11]. Third, we can distinguish between *direct* devices that enable the control of an external device through neural events that underlie, that is to say, are intrinsically related to the intended movement, and *indirect* devices that co-opt neural events that are not intrinsically related. A device that enabled the control of a robotic limb by neural signals that correlate with arm movement would be an example of a direct device; in contrast, one that controlled this movement through the suppression of cortical rhythms, or by the detection of amplitude differences between attended and non-attended stimuli, would be an indirect device.

An alternative way to categorize BCIs is to distinguish among *active*, *reactive*, and *passive* devices:

> In *active* BCIs, the user intentionally performs a mental task that produces a certain pattern of brain activity, which the BCI system detects for processing. A commonly deployed mental strategy in active BCIs is motor imagery. The user imagines moving parts of her body, without actually performing the movement. The imagination of the movement of different body parts corresponds to different activations of the primary somatosensory and motor cortical areas.

In a *reactive* BCI, brain activity is modulated in reaction to an external stimulus given by the BCI system…A commonly used paradigm is P300-based selection, where stimuli, such as letter or symbols, flash in succession on a screen. The user has to direct her attention to the symbol she wants to select. The BCI system detects the so-called P300 signal that set in 300 ms after the stimulus attended to is presented.

Passive BCIs simply monitor brain activity of the user, without requiring her to perform any mental task. The brain activity that passive systems monitor is not modulated intentionally to achieve a certain goal. ([7], pp. 3–4)

2.3   Basic and Non-basic Actions

Generally speaking, bodily movement is something that a person can just do—I do not have to think about raising my arm in order to raise it, or to think about putting one foot in front of the other in order to walk. For some people, however, bodily movement is not something that can be "just done," as the case of IW illustrates [12]. IW suffers from acute sensory neuropathy and lacks proprioceptive awareness of his body from below the neck. Despite the lack of proprioceptive awareness IW is able to control his body through visual feedback and by consciously attending to the position of his limbs and body. In order to move his arm, for example, he imagines the intended action and then controls movement by paying visual attention to his body's position.

IW's control of bodily movement is very different from the ordinary case. For most of us bodily movement is direct and immediate and experienced in a first-person way "from the inside." IW's control of bodily movement is more akin to "remote control" [13]. (Despite the challenges IW is able to lead a successful life.) Although IW's control of bodily movement is functionally and phenomenologically very different from ordinary behavior, there is good reason to claim that his bodily movement is causally related to his beliefs, and desires. A parallel conclusion seems justified in the case of LC. For in both cases sensory information is limited to visual feedback and physical movement is effected by conscious control. In LC's case we can say, therefore, that BCI-mediated behavior might be very different from ordinary behavior, but it still qualifies as intentional action.

The difficulty we face is that in the cases of IW and LC the causal process does not appear to be of the right type as defined previously. If the right type of cause is the neural activity that realizes the person's movement intention, for example, to raise an arm, then this would rule out both cases as examples of intentional action. For the proximate cause is either concentration on the movement rather than intention, or on a symbol that is used to bring movement about. In simple terms, the neural activity is of a different type.

There are a number of ways that we can respond to this difficulty. First, it could be argued that cases like those of IW and LC are sufficiently different (or rare) as to require a different analysis of action. Generally speaking, the skin-and-skull boundary defines the line between where the person begins and the world ends. Since in these cases the definition of the boundary is unclear, and the functional and phenomenological difference substantial, we could regard these cases as exceptional. It might be objected that this response does not go far enough for it suggests that our current understanding of intentional action may be too narrow. If it is the case that bodily movements can be "multiply realized," then it seems that we have to reconsider exactly what is required of the causal process to be of the right type. Critics of our current theory of action will argue, therefore, that cases like those of IW and LC reveal the limitations of our current theory, and thus warrant a revision of our present understanding of action.

An alternative approach is to say that BCI-mediated behavior is consistent with our current causal theory of action but that there are significant differences between this type of behavior and ordinary action [13, 14]. For example, it has been argued that present BCI-mediated behavior is limited to non-basic action [7]. A basic action is one that a person can perform immediately and that does not require the performance of a prior action. For example, in order to take a step forward or raise my arm I generally do not have to perform a preliminary action, nor think about the action in order to perform it. In contrast, a non-basic action requires a prior action to be performed: in order to heat the water, I have to turn on the kettle, or differently, a person learning how to play the piano has to think about where to place their hands on the keyboard. We are usually able to control bodily movement immediately and effortlessly, although it is possible to do so with deliberate conscious control. Importantly, there seems little reason to claim that the person has not acted in this latter case.

In order to effect bodily movement, the BCI-user has to perform a prior action. In the case of an indirect BCI, the user has to attend to a stimulus or consciously control brain activity and cannot just bring bodily movement about immediately. Since a direct BCI detects brain activity that is intrinsically related to the intended movement, effecting direct BCI-mediated behavior is more akin to ordinary behavior. As the user becomes experienced in controlling bodily movement the control may appear direct and immediate; but in the learning phase a high degree of conscious control is required.

Since in a number of cases the BCI-user has to consciously think about the intended movement in order to bring the movement about, whereas in ordinary bodily movement the movement is direct and immediate, the distinction between basic and non-basic action appears to point to an important difference between these types of behavior. In order to determine whether this is correct, we need to focus on two issues: first, we need to examine the distinction between basic and non-basic action; second, we need to determine whether BCI-mediated behavior qualifies only as non-basic action.

According to our causal theory, actions are distinguished from mere happenings in terms of their etiology. The raising of an arm, for example, counts as an action only if it is brought about by the relevant beliefs and desires. In other words, there is nothing about the bodily movement in and by itself that identifies it as an action rather than a mere happening. Accordingly, every bodily movement that in part constitutes an action has a necessary antecedent cause. This implies that to say a bodily movement "just happens" is not to say that the person did not think about performing the action, for otherwise the movement would be a mere happening. To say that movement "just happens" is to say that the thought process is unconscious. Accordingly, all physical actions by definition are non-basic actions—every bodily movement that is part of an action must be preceded and brought about by the agent's appropriate beliefs and desires. (If this is correct, then this raises a problem of regress for the causal theory.) I may not have to think about reaching out my hand to pick up my glass, but if this movement is to count as a physical action, rather than a mere happening, it must be causally related to my intentions. To say, therefore, that a bodily movement is direct and immediate means that the person does not have to *consciously* think about the movement in order to bring it about. In this regard, the difference between the novice and the experienced piano player is that the experienced player does not have to *consciously* think about where to place their hands.

If the analysis above is correct and all actions are non-basic actions, then BCI-mediated behavior cannot be distinguished from ordinary behavior on the grounds that it qualifies only as non-basic action. Instead, we should claim that present BCI-mediated behavior requires that antecedent mental states are conscious. It may be correct to claim that BCI-mediated behavior qualifies only as non-basic action, but this would fail to differentiate this type of behavior from ordinary behavior.

As further development of BCI technology continues, and as the user becomes more accustomed to controlling behavior in this way, we can imagine that control of the device will become more direct and immediate. Alternatively, BCI technology may provide us with a broader conception of action which recognizes that action can be realized in a variety of different ways, and the control of bodily movement need not be immediate and direct. In the future, perhaps, BCI-mediated behavior will become functionally more advanced even though phenomenologically it is brought about by "remote control."

2.4   Action, Belief, and Reliability

Our reason for concluding that Bob did not intentionally shoot the sheriff is that the proximate cause of the trigger-pulling—his nervousness—was a "mere happening" that just happened to match his original intention. It is likely that after the gun goes off Bob will say, "I didn't mean to do that!" A different scenario is that Bob fully intended to shoot the sheriff and used his predicted nervousness as the means to pull the trigger. In this case the nervousness is not just a "mere happening"—it is a reliable and intended consequence of his intention, an intended part of the causal process.

In order for the second scenario to occur certain conditions have to be met. First, there has to be a predictable and reliable causal chain that connects Bob's intention, his nervousness, and the trigger-pulling. Furthermore, Bob has to believe that such a causal chain exists. If such a connection is absent it makes little sense for Bob to form the intention to make himself nervous. Accordingly, at the price of rationality, a person has an intention to perform and action only if they believe that having this intention will bring the action about; in other words, intention requires a belief about the reliability of the connection.

> However, in order to make sense of these different types of events and states into which an agential system can enter, more than a causal relationship among particular events and states is required. This extra requirement is that the relevant causal connections among the particular events and states of the system are reliable enough to establish a distinctive type of event of state that is exclusively related to another distinctive type of event of state. ([6], p. 98)

We might say, therefore, that a particular type of mental event, for example, the intention to raise one's arm, is an event of this type because it reliably leads to the raising of the arm. Conversely, if the prior event lacked this reliability, it is difficult to see how or why we would classify the prior event as an intention-to-raise-arm event. If we now return to BCI-mediated behavior, then this implies that in order for such behavior to qualify as intentional physical action, there must exist a reliable connection that identifies the mental event as an event of a particular type, and, therefore, which explains the user forming that intention.

A second condition that needs to be met is that the bodily movement is a function of the mental state that brings the specific bodily movement about. In broad terms, we can understand the notion of function here to mean that the movement is the reliable and predictable result of the specific content of that mental state. This functional relationship allows us to explain the occurrence of the bodily movement in terms of the occurrence of the mental state.

> Now for differential explanation in nondeviant action chains, a bodily movement of a kind believed to be a ϕ-ing is differentially explained (under its bodily movement description) by the neurophysiological state that realizes for the given agent at the given time his psychological state of intending to make a bodily movement of a kind he takes to be a ϕ-ing in the circumstances (for a specific substation of ϕ). ([15], p. 70)

In order to understand Peacocke's notion of differential explanation, it is helpful to return to Bob and the Sheriff. Since the trigger-pulling is causally dependent upon a mere happening, the bodily movement is *not* differentially explained by the neurophysiological state that realizes his intention to make a bodily movement of a type he takes to be a shooting; rather, the movement is explained at least in part by the neurophysiological state that realizes his nervousness. Although the trigger-pulling is causally related to Bob's intentions, the neurophysical state that underlies the proximate cause bears no direct relation to this intention. It is for precisely this reason that differential explanation fails. As Peacocke states

> [The]...physiological states with which the nervousness interacts to produce a bodily movement that may just happen to match his original intention, but which is no function of the original intention (its neurophysiological realization) according to the principles of explanation involved. ([15], p. 70)

Let us now consider whether the two conditions presented above are met in the case of BCI-mediated behavior. To help in our discussion let us recall the case of LC who suffered a spinal cord injury and who controls the movement of a robotic limb with the aid of a BCI. Our first concern is to determine whether the connection between LC's movement intentions and the effected bodily movement is sufficient for the epistemic condition to be met. If we consider the period during which LC is learning to control the robotic movement through the BCI, it is difficult to see why LC would have the appropriate belief. In simple terms, the experience is entirely novel, and LC has no justification for believing that the movement of the robotic limb can be effected by thinking about the movement. (Analogously, if I do not know how to swim, I have no justification for believing that moving my arms and legs will propel me forward.) LC has no reason to believe that the mental state "thinking about arm movement" will bring that movement about. The causal connection is not sufficiently reliable to establish an exclusive relation between the different types of events and, therefore, to provide the basis for LC's belief.

To turn now to the second condition, this condition is met if we can differentially explain LC's movement in terms of the neurophysical states that underlie his movement intention. As described above, a distinction can be drawn between direct and indirect BCIs: in the former case the device detects and decodes information that directly correlates with the intended movement; in the latter case movement is effected by neural activity that is co-opted. To take indirect devices first, it seems plausible to claim that differential explanation is not possible precisely because the device is an indirect device: the movement is brought about (at least in part) by a neurophysical process that is distinct from that which underlies the movement intention. Accordingly, we are limited to explaining the connection in transitive terms or as a non-basic action.

In the case of direct devices, there is greater justification for claiming that the movement is effected by the appropriate neurophysical states, since the device detects and decodes LC's movement intentions from their neurophysical realizers. To assess the merit of this claim, it is important to consider LC's belief about the effected bodily movement, and in particular, whether LC takes the bodily movement to be of the appropriate kind. We might ask, for example, whether LC believes the movement of the robotic arm to be an arm-reaching. If BCIs provided the user with bodily movement that is functionally and phenomenologically similar to ordinary behavior, then there is good reason to suppose that the user will believe the movement to be of the right type (in fact, it would be surprising if the user believed otherwise). Conversely, if the effected movement is substantially dissimilar, then it is likely that the user will have a very different belief. Let us imagine that LC is learning to control bodily movement through an indirect BCI that provides minimal sensory feedback. Since the effected movement is so very different in both how it is brought about and how it is experienced by LC, it seems unlikely that LC will believe the movement to be arm-reaching. Furthermore, if LC does not believe the movement to be an arm-reaching, then the movement is not brought about by the neurophysical state that underlies LC's intention to reach out his arm. Accordingly, we cannot differentially explain the bodily movement in terms of this intention.

Since BCIs are a "novel neurotechnology," they enable the control of bodily movement (whether biological or robotic) in novel ways. We should not be surprised therefore that they challenge our existing concepts of agency and action. One way that we might defend the view that BCI-mediated behavior qualifies as intentional action is to regard the brain and BCI as a novel functional realizer or vehicle of the user's psychological states. Here we might appeal to the same line of argument as articulated by proponents of the Extended Mind (EM) theory, and argue that intention, like belief, can be realized by a coupled system that brings mind and world together [16, 17]. A full discussion of this line of argument is beyond the scope of this paper, but there are a number of considerations that militate against this approach. First, whereas it is coherent to regard belief as a dispositional state the matter is less clear with intention. It is true to say that a person can have the unconscious intention to act in some way but intending plays a more active role than the information-storage nature of belief. To say that "Smith intends to fire Jones" is to describe a predicted course of action, rather than to say that the intention is in Smith's head and is easily accessible. Second, it is not clear in the case of BCI-mediated behavior that any *cognitive* (i.e., not sensorimotor) processes are being "off-loaded." The cognitive aspects of the behavior remain "in the head."

> Motor imagery could be decoded from these populations [neurons in the posterior parietal cortex PPC], including imagined goals, trajectories, and types of movement. These findings indicate that the PPC of humans represents high-level, cognitive aspects of action and that the PPC can be a rich source for cognitive control signals for neural prosthetics that assist paralyzed patients. ([2], p. 907)

Although BCIs enable bodily movement by decoding the user's "movement intentions" it would be inaccurate to say that they decode the particular neurophysical states that underlie intentions understood in the more limited psychological sense. Rather, the device detects sub-personal level information pertaining to motor parameters like arm position, velocity, and joint torque and neural signals of imagined movements that have been detected prior to the transmission of information to the motor cortex.

It might be objected that essentially there is little difference between BCI-mediated and ordinary behavior in this regard. In both cases, there is a causal chain that connects the person's personal-level psychological states with bodily movement. If we grant, for the sake of argument, that the BCI does not detect and decode the personal-level psychological states, it can still be argued that the only difference between the types of behavior occurs "downstream": in principle, the BCI is playing the same functional role as the damaged parts of the brain and CNS, and since these damaged parts are sub-personal there is no phenomenological element to take into account. There are two responses to this last claim. First, we may well concur that there is no essential difference between the types of behavior: a BCI system that is functionally and phenomenologically like ordinary behavior should not be differentiated from ordinary behavior. Second, if such a system is developed then the user is justified in having similar beliefs in this case as in ordinary behavior; present devices, however, are considerably more limited.

2.5   Action, BCIs, and Identification

The perspective and experience of the BCI-user plays an integral role in determining whether BCI-mediated behavior qualifies as intentional action. If BCI-mediated behavior is phenomenologically very similar to ordinary behavior, then it is likely that the user will feel as though they are acting when controlling bodily movement. Conversely, if the effected movement is substantially dissimilar to ordinary movement, it is likely that the user will not perceive themselves to be in control. Since present BCI-mediated behavior is functionally and phenomenologically very different from ordinary behavior, it seems reasonable to suppose that the present BCI-user will not believe the effected movement to be of the right type. The device provides the user with only a limited sense of control, and the limited functionality and reliability of BCIs suggests that the causal process is not of the right type for the causal theory of action.

The causal theory of action typically understands actions as events, as spatiotemporal particulars. An action is differentiated from a mere happening because it is an event of a different type—a bodily movement caused by the neurophysical state that realizes the belief/desire pair. In contrast, an agent causal account regards the agent as primary—actions are what *agents do*. If we have questions about an action they are likely to be questions about what the *person* intended, rather than questions about the occurrence of a particular type of event—in asking whether the movement of BK's arm is intentional we are asking a question about *BK's* intentions, rather than about the occurrence of a particular type of brain activity.

Agent causation provides us with an alternative approach to BCI-mediated behavior. One way that we might understand freedom of action is terms of the identification by the agent and others with the action [18]. Although the BCI provides LC with limited control of movement and minimal sensory feedback, and requires his conscious control, LC may, nevertheless, come to identify with the movement. For by controlling the robotic arm LC is able to engage with the world more extensively and in a way that is causally related to intention. Accordingly, the BCI provides LC with an increased sense of autonomy. Furthermore, this increased sense of autonomy is consistent with the lack of belief that the robotic limb counts as an arm movement. Controlling the robotic limb may be functionally and phenomenologically very different from ordinary movement; nevertheless, LC may feel an increased sense of autonomy and, despite this difference, even view the robotic limb as part of the body. In this regard, BCI-mediated behavior may qualify as intentional action even though it is quite different from ordinary behavior. Controlling movement may be like "remote control" and the user may not experience this movement "from the inside" due to the lack of sensory and proprioceptive feedback. Despite these differences, the BCI-user may feel as though they are in control—that they are an agent.

For a person to intend to bring about a bodily movement, there must be a reliable connection between the intention and the movement, and the person must believe that such a connection exists. In order to intend to kick the ball, I must believe that having this intention will bring the ball-kicking about, otherwise it makes little sense to form the intention. If we apply this analysis to BCI-mediated behavior then we can say, for example, that in order to control the robotic limb LC must believe that a reliable connection exists between the intention to control and the effected movement—in simple terms, LC must believe that if the intention is formed then the robotic limb will move. Determining the conditions according to which a belief is justified is beyond the scope of this chapter; however, it is plausible to contend that it is not necessary that LC be certain that movement will occur, nor that LC has any particular beliefs about how the movement is brought about. Furthermore, if LC has this belief the question of whether BCI-mediated behavior is a basic or non-basic action becomes unimportant. In order to intend bodily movement LC must believe that having the intention brings the movement about. This is consistent with LC believing that thinking about the movement can bring the movement about (non-basic) and that the movement "just occurs" (basic).

A final point about the identification of action relates to the perceptions of others. Unlike mental actions, physical actions are public events and are defined by their context. Accordingly, our interpretation of the movement of LC's robotic arm will be based in part on our familiarity with the technology and context. Thus, for better or for worse, the more the movement is consistent with ordinary behavior, the more we are likely to regard the movement as an action, and not as a mere happening.

2.6   Conclusion

According to a common understanding, actions are bodily movements that are causally related to a person's beliefs and desires. If the analysis in this chapter has been successful, then there is reason to suggest that explicating this causal relation is not a straightforward matter. Progress in the development of BCI technology in recent years has been remarkable, and in the future BCIs may provide the user with a control of bodily movement that is just like ordinary behavior. The main conclusion of this chapter is that this level of functional and phenomenological similarity is not a requirement for BCI-mediated behavior to qualify as intentional action.

References

1. Bosely S. Paralyzed man moves arm by power of thought in world first. The Guardian. 2017. https://www.theguardian.com/science/2017/mar/28/neuroprosthetic-tetraplegic-man-control-hand-with-thought-bill-kochevar.
2. Aflalo T, et al. Decoding motor imagery for the posterior parietal cortex of a tetraplegic human being. Science. 2015;348(6237):906–10.
3. Carmena JM. Advances in neuroprosthetic learning and control. PLoS Biol. 2013;11(5):e1001561. https://doi.org/10.1371/journal.pbio.1001561.
4. Patil PG, Turner DA. The development of brain-machine interface neuroprosthetic devices. Neurotherapeutics. 2008;5:137–46.
5. Bates M. Converting thoughts into action. BrainFactsorg. 2018. Accessed 16 Jul 2019.
6. Aguilar JH, Buckareff AA. Causing human actions: new perspectives on the causal theory of action. Cambridge: MIT Press; 2010.

7. Steinert S, et al. Doing things with thoughts: brain-computer interfaces and disembodied agency. Philos Technol. 2019;32:457–82. https://doi.org/10.1007/s13347-018-0308-4.
8. Carmena JM, et al. Learning to control a brain-machine interface for reaching and grasping by primates. PLoS Biol. 2003;1(2):e42. https://doi.org/10.1371/journal.pbio.0000042.7.
9. Collinger JL, et al. High-performance neuroprosthetic control by an individual with tetraplegia. Lancet. 2013;381(9866):557–64.
10. Donoghue JP. Bridging the brain to the world: a perspective on neural interface systems. Neuron. 2008;60(3):511–21.
11. Lebedev MA, Nicolelis MA. Brain-machine interfaces: past, present, and future. Trends Neurosci. 2006;29(9):536–46.
12. Gallagher S. How the body shapes the mind. Oxford: Oxford University Press; 2005.
13. Wong HY. Embodied agency. Philos Phenomenol Res. 2017;97(3):584–612. https://doi.org/10.1111/phpr.13292.
14. Wong HY. On the significance of bodily awareness for bodily action. Philos Q. 2015;65(261):790–811.
15. Peacocke P. Holistic explanation: action, space, interpretation. Oxford: Clarendon Press; 1979.
16. Clark A, Chalmers D. The extended mind. Analysis. 1998;58:7–19.
17. Clark A. Re-inventing ourselves: the plasticity of embodiment, sensing, and mind. J Med Philos. 2007;32:263–82.
18. Farahany NA. A neurological foundation for freedom. Stanford Technol L Rev. 2012;4:1–15.

Skilled Action and the Ethics of Brain-Computer Interfaces

Sebastian Drosselmeier and Stephan Sellmaier

Contents

Abstract

By creating an artificial output path for the central nervous system, Brain-Computer Interfaces (BCIs) enable users to affect the world without engaging their bodies' peripheral muscular and nervous systems. The action-theoretical as well as normative implications of this novel way of acting have recently gained attention in philosophical discourse. What, if anything, are the differences between ordinary bodily actions and actions mediated by machines or computers? And what are the normative implications that result from these differences?

S. Drosselmeier (✉)
Munich Graduate School for Ethics in Practice, Ludwig-Maximilians-Universität (LMU)
München, Munich, Germany
e-mail: s.drosselmeier@lrz.uni-muenchen.de

S. Sellmaier
Research Center for Neurophilosophy and Ethics of Neurosciences, Ludwig-Maximilians-Universität (LMU) München, Munich, Germany
e-mail: sellmaier@lmu.de

O. Friedrich et al. (eds.), *Clinical Neurotechnology meets Artificial Intelligence*,
Advances in Neuroethics, https://doi.org/10.1007/978-3-030-64590-8_3

We argue for a new focus in addressing these questions by highlighting an action-theoretical aspect that has been neglected so far: the acquisition of skill in acting with BCIs. By drawing on empirical literature, we show that skilled BCI users are able to perform actions with their devices without performing a mental act. This results in an action-theoretical analysis of BCI actions according to which users are able to perform BCI actions as basic actions. We then discuss some normative implications of our findings, especially for the applicability of the legal concept of actions to actions performed using BCIs.

3.1   Introduction

Brain-Computer Interfaces (BCIs), also sometimes referred to as Brain-Machine Interfaces (BMIs), enable users to act and affect the world in hitherto unprecedented ways. By connecting the user's brain directly to a computer, BCIs generate an artificial output path for acting that circumvents the peripheral muscular and nervous system and allows users to control a machine or a computer program. Hence, in a yet to be specified way, BCIs enable users to *do things with thoughts*—and without the engagement of muscles or nerves.

BCIs possess vast potential for medical use [1, 2] and other non-medical purposes, such as media applications and gaming [3]. This disruptive potential has been widely recognized in the academic literature as well as in business.[1] In addition, BCIs pose a variety of philosophical questions. These range from the conceptual analysis of actions performed by using BCIs to normative worries about the further implications of the technology. One of the most fundamental challenges posed by the development of BCIs concerns the categorical differences between ordinary (i.e., bodily-mediated) actions and novel (i.e., BCI-mediated) actions. What, if anything, is the difference between raising my arm to grasp a cup of coffee and controlling a robotic arm through a BCI to do the same thing? And does such a difference have any normative implications?

Some authors have suggested that BCI-mediated actions possess a variety of peculiar features that categorically differentiate them from ordinary, bodily actions and implicate particular ethical and legal challenges [4, 5]. The alleged idiosyncrasy of BCI-mediated actions centers around the idea that in using BCIs, people do not just do things directly *with* their thoughts but rather indirectly *by* their thoughts. This instrumental and—figuratively speaking—*remote* role of mental actions in relation to the activity realized by a machine or a computer, the authors argue, makes it difficult to apply both philosophical and legal theories of action to BCIs. In the next section, we will outline this approach in detail before then moving on to empirically and philosophically criticizing it. Ultimately, it is the aim of our paper

---

[1] While Mark Zuckerberg's *Facebook* is working on a BCI to enable users to generate text directly from thoughts, Elon Musk's *Neuralink*—in the long run—aims at nothing less than mankind's merging with artificial intelligence.

to highlight a dimension of human action that has so far been neglected in both action-theoretical debates as well as normative debates surrounding BCIs, namely the level of proficiency or *skill* in using the technology. We will argue for the conditional claim that when a user acquires sufficient skills in the use of a BCI, BCI-mediated actions do not differ from ordinary actions in any relevant ways. Hence, just as raising one's arm is a paradigmatic example of an act, so is raising one's BCI-controlled prosthetic arm.

We will start by outlining the rationale for differentiating between traditional actions and BCI actions, which is based on the distinction between basic and non-basic actions in different theories of action. By relating this distinction to the use of BCIs, we will highlight the features in virtue of which categorical differences between the two types of actions are commonly drawn and what seems to be peculiar about BCI actions (cf. [4]). We will then introduce the general notion of *skilled action* and argue that if an agent reaches a sufficient level of proficiency, the conceptual border between basic and non-basic actions dissolves. We will support the thesis that BCI users can become skilled in using their devices by drawing on empirical literature related to BCI training and neuroplasticity. Lastly, we will discuss the ethical and legal implications of our findings and argue that the skilled use of BCIs should be seen as the paradigmatic case of BCI-mediated actions. Hence, although the supposed peculiarities of acting with BCIs are not fully unwarranted, they are limited to the borderline cases of acquiring the skills necessary to become a competent user of a BCI.

3.2 Do BCI Actions Differ from Ordinary Actions?

3.2.1 How Does Acting with a BCI Work?

Before diving into the action-theoretical aspects of BCIs, a brief introduction into the way they function as well as a corresponding typology of different BCIs is due. This will help us make the theoretical viewpoint that we will be concerned with throughout this paper plausible. Authors should explicitly mention which part of the developmental phase they are interested in when philosophically investigating a scientific or technological phenomenon. In this paper, we will focus exclusively on BCIs that have reached a certain level of technological maturity and hence, are not far from becoming a market-ready product. It is reasonable to assume that such a product would not exhibit major design flaws nor include technological aspects which—although in principle conceivable—would turn out to be undesirable from the perspective of its users.[2] Instead, these products would fulfil the main objective

---

[2] Take the well-known example of Ichiro, whose behavior is controlled by a (possibly ill-intentioned) neuroscientist via a brain implant ([6], p. 87). Such cases are important for the action-theoretical debate of deviant causal chains (see for example [7], pp. 10–11; [8]). However, they are irrelevant for the purpose of this paper, since such a device would not allow users to exercise voluntary control.

of BCI research: restore lost or deficient motor control and enable users to voluntarily control prosthetic limbs, computer programs, or the like.

Let us then start with the way BCIs function. As mentioned in the introduction, BCIs enable users to control a computer or a machine without the engagement of the peripheral muscular or nervous systems. Instead, it is often said that BCIs enable users to *do things with their thoughts* (see for example [4, 9–11]). However, the focus on the broad concept of *thoughts* is imprecise and can lead to vast misunderstandings (cf. [12], p. 30). More precisely speaking, BCIs rely on the measurement of electrical or magnetic brain activity ([2], p. 6). The respective brain activity can be accessed non-invasively, i.e., without surgery, for example by electroencephalography (EEG), or invasively, i.e., with surgery, for example by recording the electrocorticogram (ECoG) ([2], pp. 7–8).[3] Brain activity accessed in these ways correlates with mental activity and is computationally interpreted to control a machine or a computer program. However, BCIs are not able to "read minds"—if one believes that mind reading consists of, for example, accessing the introspective and subjective sphere of human experience. While we do believe that this is common ground in the BCI community it is still worth highlighting, in that much of the action-theoretical debate and subsequent normative worries—somewhat surprisingly—identify the role of thoughts as one of the main peculiarities of acting with BCIs ([13], pp. 388–389). We will treat this point in greater detail below, but for now it is important to highlight that BCIs react to physically realized brain activity and not to thoughts *qua thoughts*.

With a basic understanding of the way BCIs function, we can now additionally delineate *active*, *reactive,* and *passive* BCIs ([14], pp. 74–77). To control an active BCI, users have to produce a certain pattern of brain activity, which is recorded and translated by a specifically attuned decoder to allow the control of a machine or computer program. In this way it is possible to enable users to control prosthetic limbs, wheelchairs, or computer-cursors (cf. [4], p. 5). To control a reactive BCI, on the other hand, users do not need to produce specific patterns of brain activity, but instead merely attend to externally presented stimuli. For example, users can be enabled to write a text by focusing on letters appearing on a screen. Reactive BCIs mostly function by measuring the P300 event related readiness potential that allows the respectively attuned decoder to determine which object the user has attended to and hence generate the corresponding letter on a screen (cf. [4], p. 5).

Lastly, a third category of passive BCIs has been proposed ([14], p. 76; [4], p. 5). This third category of BCIs does not require users to build up specific brain patterns or attend to external stimuli, but instead reacts to brain activities that are associated with phenomena beyond their intentional control, such as workload or arousal. Yet some authors have suggested that devices that merely react passively and do not allow the user to exercise voluntary control should not be classified as BCIs ([2], p. 3). Even more importantly, Steinert et al. ([4], p. 12) conclude that events mediated by passive BCIs do not count as actions in a standard action-theoretical sense,

---

[3] The details of these different technologies are beyond the scope of this paper. However, we will come back to some of the technical details in Sect. 3.3.

since such events are not realized by the relevant mental states of BCI users. Since we are interested in action-theoretical and related normative implications of the use of BCIs, so-called passive BCIs will be disregarded for the remainder of this paper.

In sum, acting with a BCI involves three components: (a) mental activity, (b) the corresponding physically realized brain activity, and (c) whatever it is the machine eventually does in reaction to these. How these three components interact with each other will turn out to be crucial for an adequate action-theoretical analysis of BCI actions. Yet, before developing our own account, we want to clarify the foil that we want to compare it to. As mentioned earlier, some authors argued that BCI actions exhibit peculiar features that categorically differentiate them from ordinary actions [4]. This difference can be explained by the role ascribed to mental activity in conjunction with the concept of *basic actions*, which we will turn to now.

3.2.2   What Is Peculiar About BCIs?

What are basic and non-basic actions? The difference between the two types of actions can be illustrated in terms of *instrumentality*. While a basic action is an immediate action, the performance of which does not require further, preceding actions, the performance of a non-basic action does require doing something else first [15]. For example, one can raise one's arm *in order to* give a signal to one's confederates ([16], p. 525). Here, the immediate raising of one's arm serves as a basic action, while the signal thereby given serves as a non-basic action. In addition to the distinction between basic and non-basic actions, Steinert et al. [4] differentiate *mental* from *bodily* actions. Following Mele [17], they define mental actions as actions not involving bodily movements, such as calculating the result of some numerical exercise in your head. Bodily actions, on the other hand, involve and require motor outputs, such as the lifting of your hand ([4], pp. 8–9).

Now, according to Steinert et al., while the lifting of one's arm is a paradigmatic instantiation of a basic and bodily action, the two types of distinctions yield a peculiar result for BCI-mediated actions: since BCIs read off the user's brain states, the user's basic action is a mental act which then causes the non-basic act of, for example, the movement of a robotic arm. Hence, "[t]he BCI-mediated event in *active* and *reactive* BCIs is a *non-basic* action [...] because the subject always performs it *by* doing something else" ([4], p. 10). And

> [t]hus, an action with identical effects—say, an arm reaching for a glass—is a basic action under natural circumstances (that is in non-BCI contexts), but a non-basic action when it is facilitated with a BCI-prosthesis ([4], p. 10).

What exactly is it that users have to do in order to perform the—on this view—non-basic act of reaching for a glass with a BCI-prosthesis? Steinert et al. argue that users need to give a "go-command" which further distinguishes BCI actions from ordinary actions ([4], p. 13). While in a normal context people are able to simply grasp the glass, BCI users need to first *activate* their BCI, which then in turn performs the desired task. This activation is achieved by consciously thinking a specific thought, which is read out by the machine and translated into the corresponding action.

This picture of acting with BCIs has severe normative implications, especially in the legal realm. Firstly, Steinert et al. argue that the current concept of actions in the law excludes most BCI actions ([4], p. 18). This is due to the fact that BCI actions do not involve bodily movements.[4] Since the act performed by the agent is merely a mental act, while the movements thereby initiated affect the world as non-basic acts, the legal concept of an action does not apply ([4], p. 16). This leads to a further normative peculiarity, namely that law makers would have to punish BCI users for their thoughts if they committed a crime via their BCIs ([4], p. 21). Hence, BCI users would be prohibited from thinking certain thoughts that trigger the respective unlawful tasks executed by their BCIs. Thus, the authors conclude, assuming that certain thoughts are legally prohibited, "regular BCI users will (have to) develop novel skills to watch over their own mind" ([4], p. 21).

In sum on the view that was just introduced, the interplay between thoughts, brain activity, and the subsequent operation of a machine leads to a categorical distinction between bodily-mediated and BCI-mediated actions. While bodily actions are basic, BCI actions are non-basic. In the remainder of this paper, we will argue that this analysis of BCI actions only enjoys prima facie plausibility when focusing on inexperienced BCI users. The action-theoretical analysis of experienced BCI users, on the other hand, will turn out to be substantially different. Hence, we will now attempt to complete the action-theoretical analysis of BCIs.

3.3   Is Acting with a BCI a Skill Users Can Acquire?

3.3.1   What Is Skilled Acting?

Let us begin with a very general observation of human actions: by acquiring a skill, one and the same action and bodily movement can be realized in different ways—either automatically or with full attention. It is this very human ability that allows us to perform complex actions. As competent drivers, we can all have a demanding conversation with our passenger, requiring our full attention, without losing control of our vehicle. We can do this because as competent drivers we have acquired the skill to drive our car subconsciously or automatically. This does not mean that we have given up control over our car, but only that driving a car does not constantly need our full consciousness/attention. Only in critical, challenging situations do we need to focus our attention on driving. This enables us to perform complex, multilayered actions such as driving and talking or listening and writing simultaneously and in parallel. These skills, however, are not innate to us, but are owed to the

---

[4]This view is controversial. Whether BCI actions involve bodily movement is a question that has been addressed in the broader discussion of the causal theory of action—independent from any legal concerns ([18], pp. 31–32). We will come back to this point towards the end of this paper.

laborious process of learning different ways of acting—such as driving. The beginner would do well to devote his full attention to driving. Only with a certain level of skill will it gradually be possible to subconsciously control the various aspects of driving that are coordinated with each other. More and more often she will no longer change gears intentionally and with full attention, but will simply start to shift gears automatically when necessary. Shifting will become—as the saying goes—her second nature. This rather general observation about human action is important for the philosophical theory of action insofar as it shows us that the basic model of intentional action controlled with full attention is just one aspect of human action. Our ability to execute even simple everyday actions in different ways—such as switching on the light—distinguishes and enables us to perform complex, compound actions.

If we assume this general picture of human action, the following question becomes pertinent. To what extent does using BCIs permit such subconscious, automated and skilled actions, or do the partially outsourced technical properties of BCIs completely exclude this possibility? In other words, is it possible to become a *skilled* user of BCIs? Interestingly, the notion of skilled BCI use has been widely discussed in the neuroscientific literature. This is unsurprising, since investigating what it takes to acquire a skill in using BCIs permits engineers to create a more natural and seamless experience for patients. However, neither the action-theoretical nor the normative implications of skilled BCI use have so far been discussed or recognized and we intend to close this research gap. The remainder of this section will therefore be dedicated to empirically supporting the notion of skilled BCI use. Additionally, we will further refine the action-theoretical understanding thereof. In the subsequent section, we will examine the normative implications emerging from our new picture of BCI actions.

3.3.2   Learning Effects and BCI Skills

There is vast support in the empirical literature for the idea that users can become competent in using BCIs. For example, Koralek, Jin, Long II, Costa & Carmena [19] write that

> [t]he ability to learn new skills and perfect them with practice applies not only to physical skills but also to abstract skills, like motor planning or neuroprosthetic actions,

and Wander et al. [20] conclude that

> [t]he ability to voluntarily modulate neural activity to control a brain–computer interface (BCI) appears to be a learned skill, similar to learning to ride a bike or swing a golf club.

Acquiring this skill can be compared to natural learning [10, 21]. Learning BCI skills is possible due to the high degree of plasticity in cortical neuronal ensembles [22] and neural adaptions through training [20, 23]. Ultimately, a user can become competent enough in using a BCI so that performing the act in question becomes proverbial second nature and he gains a sense of ownership of the prosthesis [21, 24].

The basic psychological mechanism that underlies learning BCI skills is called operant conditioning ([2], p. 13). Operant conditioning enables people to relate the actions they perform to their external effects. A standard example is touching a hot stove and understanding that this results in experiencing pain ([2], p. 13). Applied to BCIs, the initial action people have to perform in the beginning of the learning period is the mental act in question, while the effect is the operation of the computer or machine. The process of operant conditioning relies heavily on direct feedback, which enables users to distinguish between successful and unsuccessful attempts. This is one of the reasons why providing feedback is so central to BCIs and why some authors understand real-time feedback as a necessary condition for a device to even conceptually classify as a BCI ([2], p. 3).

Learning periods and operant conditioning enhance a user's control over her device. Given the action-theoretical picture we discussed above, one might now be inclined to believe that BCI learning merely leads to a consolidation of the mental act, which triggers the machine. On this view, users would become better at performing the required mental task, for example by instantiating the required brain patterns more robustly. Yet, this is not the main effect operant conditioning has on the brain. The effect instead runs far deeper and eventually supersedes the need to perform a mental act at all. Neuper and Pfurtscheller ([25], p. 68) give a detailed account of the way neurofeedback works. They refer to the *basket game*, a computer game in which BCI users have to direct a ball into a highlighted basket via an EEG-based BCI. While players in this game usually experience a period of "trial-and-error"—or, to employ the terminology just introduced, a period of operant conditioning—in which feedback enables them to enhance their performance, the ultimate effect of learning is that

> [o]ver a number of sessions, the subject probably acquires the skill of controlling the movement of the ball without being consciously aware of how this is achieved ([25], p. 68).

In other words:

> With continued practice, [...] control tends to become automatic, as is the case with many motor skills and imagery becomes unnecessary ([26], pp. 98–99).

Thus, we can conclude that it indeed seems to be possible to act with a BCI in a skillful way, just as it is possible to drive a car while having a demanding conversation.[5] After a successful learning period, control over the BCI becomes automatic and does not require a mental act to activate the BCI.[6] In the next section, we will

---

[5] For a discussion on acting with BCIs while being distracted, see [27].

[6] It was objected that the learning effect relates to the instantiation of the required brain pattern instead of the BCI itself. However, this reinterpretation would fail to establish a difference between BCI actions and ordinary actions and lead to the implausible result that *any* skilled action is non-basic, since any such action requires the instantiation of a specific brain pattern which is learned through training.

summarize what this means for the action-theoretical analysis of BCIs before then moving on to some normative implications.

3.3.3   Skilled Acting in Action Theory

In the last two sections we introduced the notion of skill and argued that acting with a BCI is a skill that a user can learn, which is similar to learning how to walk, ride a bike, drive a car, or speak. What does this imply for the action-theoretical analysis of BCIs? Let us come back to the role of conscious thoughts or mental acts in activating BCIs. As mentioned above, overstating the role of thoughts as mental acts can lead to vast misunderstandings. These misunderstandings are not only of technological and neuroscientific nature, but also concern the action-theoretical and normative implications of BCIs. Wolpaw and Boulay ([12], p. 30) clarify the role of thoughts in BCI use in the following way:

> BCIs are not "mind-reading" or "wire-tapping" devices that listen in on the brain, detect its intent, and then accomplish that intent directly rather than through neuromuscular channels. This misconception ignores a central feature of the brain's interactions with the external world: that the skills that accomplish a person's intent, whether it be to walk in a specific direction, speak specific words, or play a particular piece on the piano, are mastered and maintained by initial and continuing adaptive changes in central nervous system (CNS) function.

These changes in the CNS enable users to perform BCI actions without performing a conscious mental act. Users can control their devices automatically, even under light distraction, and—crucially—without the performance of an initial, basic act. It follows then that the action-theoretical approach presented in the previous section has to be revised. When BCI users acquire the skill in question, they no longer need to attend to the cognitive task, i.e., think a specific thought to generate the corresponding brain activity, but instead directly control their BCI.

Hence, the distinction between basic and non-basic action that lies at the core of the argument presented by [4] breaks down. For skilled BCI users, there is no difference between raising a bodily arm and raising a BCI-controlled prosthetic arm. Both of these actions are basic in the sense that performing them does not require any preceding, instrumental actions. Note, again, that this is merely a conditional claim which holds for skilled BCI users. Those in early phases of the learning process will indeed need to produce thoughts associated with the respective brain patterns to control their devices—while being more or less successful in doing so. Yet, it is crucial to note that this is not a peculiar feature of BCI actions in general, but merely a feature of the learning phase. Once a user has consolidated his BCI skills, he no longer needs to act in this peculiar way.[7] Instead, the user is able to switch between a deliberate, attentive realization of the action and a subconscious one.

---

[7] Whether every BCI user is able to acquire these necessary skills is a question open for empirical investigation.

This also means that skilled BCI users are able to differentiate between entertaining a thought, say the wish to move an arm, and its actual realization. With these insights at hand, we will now turn to clarify some normative aspects that are often associated with BCIs.

3.4   The Normative Implications of Skilled BCI Use

In the last section, we refined the action-theoretical understanding of BCIs by introducing the notion of *skill*. Let us now discuss the normative implications of our findings. To begin with, consider the general significance of the level of competence on further normative evaluations. For example, drivers who only recently obtained a driver's license and lack substantial driving experience have to pay a higher insurance premium than experienced drivers. This is because they exhibit less control over their vehicles and are thus expected to be involved in accidents more frequently. In the same manner one can expect new BCI users to be less able to exercise control over their devices. This is why extensive training periods are not only a necessity for the acquisition of the skill of acting with a BCI in a subconscious, automatic manner, but also for the ability to sufficiently control it.

The need for training also helps to illuminate a first normatively relevant issue that is often associated with BCIs, namely the lack of a "sense of agency" that some BCI users report (cf. [4], pp. 7–8). Without the acquisition of a skill, which may ultimately lead to the integration of the device into the user's body schema [22], users will indeed perceive the device as foreign and not under their voluntary control. Yet, this too is not a phenomenon inherently related to BCIs, but merely a consequence of a lack of sufficient training. It then seems adequate not to classify new users without sufficient training as paradigmatic examples for philosophical analysis, but to instead focus on more experienced users.

The notion of skilled BCI use allows us, secondly, to clarify an aspect in the legal debate mentioned in Sect. 3.2.2. Since skilled BCI users do not need to perform a mental act to control their devices but can instead do so subconsciously, makes the worry that users would need to be punished for their thoughts unwarranted. Again, this worry seems to be rooted in a misattribution of the role of thoughts or mental acts in the way that BCIs function. In light of the empirical studies presented we argued for a different conception of BCI actions, which is shared and aptly summarized by Orsborn and Pesaran ([28], p. 77):

> Rather than "reading out" thoughts about movement, motor BMIs give the brain a new tool it learns to control by constructing a new neural representation.

We, therefore, want to propose a shift in the debate on BCIs, which is to perceive BCIs as mere tools. Yet, to make this approach plausible we need to address a common objection, which states that BCI actions do not involve bodily movements. For example, Steinert et al. ([4], p. 19) argue that BCIs cannot be understood as mere tools, since "[o]rdinary controlling of tools requires bodily movements, controlling them through BCIs does not." This aspect is of further relevance since it has also been argued that due to a lack of bodily movement BCI actions do not count as actions in the legal sense (cf. [5]). Yet, as mentioned earlier, whether BCI actions involve bodily movement is a much-debated question. Surely, what is not involved in BCI actions is the peripheral muscular system, which might in everyday speech often be equated with "bodily movement." However, it is also not the case that BCI users control their devices via their thoughts. Instead, users are able to control their devices via the measurement of changes in their brain activity. Moore therefore argues—citing Davidson's suggestion to "interpret the idea of a bodily movement generously" ([29], p. 49)— that actions facilitated by BCIs do involve bodily movements ([18], p. 32). Yet, Moore also notes that it would be a grave mistake to understand every change in the brain as a bodily movement, since only those changes that are caused with the intention to have some "effect in the world" plausibly qualify as actions ([18], p. 32). And since this is undoubtedly the case, especially for skilled BCI actions, these can be interpreted as actions in a standard legal sense.

In sum, we therefore believe that understanding BCIs as mere tools is both empirically adequate and conceptually sound. By adopting a generous understanding of bodily movement, the worry that the legal concept of an action does not apply to BCI actions vanishes, as well. Instead, with the full action-theoretical analysis at hand, the law seems to be well equipped to handle the emergence of BCIs in the legal realm.

3.5   Conclusion

The objective of this paper was to supplement the action-theoretical analysis of BCIs with the notion of *skilled action*. We showed that acting with a BCI, automatically or subconsciously, is possible due to the effects of learning and the brain's plasticity. This resulted, firstly, in a novel view of acting with BCIs. Skilled BCI users do not need to perform a mental act in order to activate their devices, but can instead exercise direct control over them. Hence, skilled BCI users can perform BCI actions as basic actions. This also means that BCI users are able to distinguish between entertaining a thought and acting on an intention. Secondly, this led us to revise some normative implications often related to BCI actions. Skilled users are able to voluntarily control their devices and hence to voluntarily bring about effects in the world. BCIs can therefore be classified as tools that users can become competent in using. This conceptualization is especially relevant for the law, since it enables the standard legal framework to be applicable to BCI actions.

References

1. Kuebler A, Kotchoubey B, Hinterberger T, Ghanayim N, Perelmouter J, Schauer M, et al. The thought translation device: a neurophysiological approach to communication in total motor paralysis. Exp Brain Res. 1999;124(2):223–32. PubMed PMID: 9928845. Epub 1999/02/03.
2. Graimann B, Allison BZ, Pfurtscheller G. Brain–computer interfaces: a gentle introduction. In: Graimann B, Allison BZ, Pfurtscheller G, editors. Brain-computer interfaces: revolutionizing human-computer interaction. Berlin: Springer Science & Business Media; 2010.

3. Blankertz B, Tangermann M, Vidaurre C, Fazli S, Sannelli C, Haufe S, et al. The Berlin brain-computer interface: non-medical uses of BCI technology. Front Neurosci. 2010;4:198. PubMed PMID: 21165175. PMCID: PMC3002462. Epub 2010/12/18.
4. Steinert S, Bublitz C, Jox R, Friedrich O. Doing things with thoughts: brain-computer interfaces and disembodied agency. Philos Technol. 2018;32(3):457–82.
5. Bublitz C, Wolkenstein A, Jox RJ, Friedrich O. Legal liabilities of BCI-users: responsibility gaps at the intersection of mind and machine? Int J Law Psychiatry. 2019;65:101399. PubMed PMID: 30449603. Epub 2018/11/20.
6. Peacocke C. Holistic explanation: action, space, interpretation. Oxford: Clarendon Press; 1979. p. 106–18.
7. Aguilar JH, Buckareff AA. The causal theory of action: origins and issues. In: Aguilar JH, Buckareff AA, editors. Causing human actions: new perspectives on the causal theory of action. Cambridge: MIT Press; 2010. p. viii, 327 p.
8. Aguilar JH. Agential systems, causal deviance, and reliability. In: Aguilar JH, Buckareff AA, editors. Causing human actions: new perspectives on the causal theory of action. Cambridge: MIT Press; 2010. p. viii, 327 p.
9. Glannon W. Ethical issues in neuroprosthetics. J Neural Eng. 2016;13(2). PubMed PMID: 26859756. Epub 2016/02/10:021002.
10. Green AM, Kalaska JF. Learning to move machines with the mind. Trends Neurosci. 2011;34(2):61–75. PubMed PMID: 21176975. Epub 2010/12/24.
11. Lee JH, Ryu J, Jolesz FA, Cho ZH, Yoo SS. Brain-machine interface via real-time fMRI: preliminary study on thought-controlled robotic arm. Neurosci Lett. 2009;450(1):1–6. PubMed PMID: 19026717. PMCID: PMC3209621. Epub 2008/11/26.
12. Wolpaw JR, Boulay CB. Brain signals for brain–computer interfaces. In: Graimann B, Allison BZ, Pfurtscheller G, editors. Brain-computer interfaces: revolutionizing human-computer interaction. Berlin: Springer Science & Business Media; 2010.
13. Bublitz C. Der rechtliche Handlungsbegriff. In: Kühler M, Rüther M, editors. Handbuch der Handlungstheorie. Stuttgart: JB Metzler; 2016.
14. Thurlings ME, van Erp JB, Brouwer A-M, Werkhoven PJ. EEG-based navigation from a human factors perspective. Brain-computer interfaces. Berlin: Springer; 2010. p. 71–86.
15. Danto AC. Basic actions. Am Philos Q. 1965;2(2):141–8.
16. Clarke R. Skilled activity and the causal theory of action. Philos Phenomenol Res. 2010;80(3):523–50.
17. Mele AR. Agency and mental action. Philos Perspect. 1997;11:231–49.
18. Moore MM. Renewed questions about the causal theory of action. In: Aguilar JH, Buckareff AA, editors. Causing human actions: new perspectives on the causal theory of action. Cambridge: MIT Press; 2010. p. viii, 327 p.
19. Koralek AC, Jin X, Long JD 2nd, Costa RM, Carmena JM. Corticostriatal plasticity is necessary for learning intentional neuroprosthetic skills. Nature. 2012;483(7389):331–5.PubMed PMID: 22388818. PMCID: PMC3477868. Epub 2012/03/06.
20. Wander JD, Blakely T, Miller KJ, Weaver KE, Johnson LA, Olson JD, et al. Distributed cortical adaptation during learning of a brain-computer interface task. Proc Natl Acad Sci U S A. 2013;110(26):10818–23. PubMed PMID: 23754426. PMCID: PMC3696802. Epub 2013/06/12.
21. Shenoy KV, Carmena JM. Combining decoder design and neural adaptation in brain-machine interfaces. Neuron. 2014;84(4):665–80. PubMed PMID: 25459407. Epub 2014/12/03.
22. Shokur S, O'Doherty JE, Winans JA, Bleuler H, Lebedev MA, Nicolelis MA. Expanding the primate body schema in sensorimotor cortex by virtual touches of an avatar. Proc Natl Acad Sci U S A. 2013;110(37):15121–6. PubMed PMID: 23980141. PMCID: PMC3773736. Epub 2013/08/28.
23. Meng J, Zhang S, Bekyo A, Olsoe J, Baxter B, He B. Noninvasive electroencephalogram based control of a robotic arm for reach and grasp tasks. Sci Rep. 2016;6:38565. PubMed PMID: 27966546. PMCID: PMC5155290. Epub 2016/12/15.

24. Collins KL, Guterstam A, Cronin J, Olson JD, Ehrsson HH, Ojemann JG. Ownership of an artificial limb induced by electrical brain stimulation. Proc Natl Acad Sci U S A. 2017;114(1):166–71. PubMed PMID: 25459407. PMCID: PMC5224395. Epub 2016/12/21.
25. Neuper C, Pfurtscheller G. Neurofeedback training for BCI control. In: Graimann B, Allison BZ, Pfurtscheller G, editors. Brain-computer interfaces: revolutionizing human-computer interaction. Berlin: Springer Science & Business Media; 2010.
26. Sellers EW, McFarland DJ, Vaughan TM, Wolpaw JR. BCIs in the laboratory and at home: the Wadsworth Research Program. In: Graimann B, Allison BZ, Pfurtscheller G, editors. Brain-computer interfaces: revolutionizing human-computer interaction. Berlin: Springer Science & Business Media; 2010.
27. Allison B. Toward ubiquitous BCIs. In: Graimann B, Allison BZ, Pfurtscheller G, editors. Brain-computer interfaces: revolutionizing human-computer interaction. Berlin: Springer Science & Business Media; 2010.
28. Orsborn AL, Pesaran B. Parsing learning in networks using brain-machine interfaces. Curr Opin Neurobiol. 2017;46:76–83. PubMed PMID: 28843838. PMCID: PMC5660637. Epub 2017/08/28.
29. Davidson D. Essays on actions and events. 2nd ed. Oxford: Oxford University Press; 1980. p. xxi, 324 pp.

Augmenting Autonomy Through Neurotechnological Intervention à la Kant: Paradox or Possibility?

4

Anna Frammartino Wilks

Contents

Abstract

The era of human-machine integration, offering a vast scope for augmenting our capacities as persons, simultaneously threatens to diminish them. This chapter examines the metaphysical, epistemological, and ethical implications for the autonomous agency of persons with respect to human-machine integration through brain-computer interfacing (BCI). The aim is to assess the possibility of integrating a person via a neurotechnological mechanism, such as BCI, for the purpose of enhancing their autonomous agency. Given the externality of the components of BCI and the need for autonomous agency to be unaffected by factors external to the self in order to constitute an authentic form of self-legislation, the objective of augmenting autonomy in this way seems hopelessly paradoxical. Operating within a broadly Kantian framework, however, there is much scope for extending the notion of the self to accommodate a form of

A. F. Wilks (✉)
Department of Philosophy, Acadia University, Wolfville, NS, Canada
e-mail: anna.wilks@acadiau.ca

© Springer Nature Switzerland AG 2021
O. Friedrich et al. (eds.), *Clinical Neurotechnology meets Artificial Intelligence*,
Advances in Neuroethics, https://doi.org/10.1007/978-3-030-64590-8_4

self-legislation that includes the function of the BCI components. Only in this way, I maintain, can the prospect of augmenting autonomy through BCI avoid the threat of paradox.

4.1   Introduction: The Basis for Augmenting Autonomy Through Brain-Computer Interfacing

Among the most valued features of persons is *autonomy*. The preservation of and respect for autonomy is, therefore, among our deepest concerns when confronted with the opportunity to enhance the lives of persons through emerging technologies. This concern is particularly pertinent in the age of human-machine integration, which presents the possibility of undermining as well as augmenting our capacities as persons. This chapter examines the metaphysical, epistemological, and ethical implications for the autonomous agency of persons with respect to human-machine integration through brain-computer interface technology (BCI). Brain-computer interfacing involves the coupling of the agent's brain with some neurotechnological devices and robotic controllers that are in a very real sense *external* to the agent. This being the case, it is not clear how it may reasonably be said that the agent, in deliberating and acting in accordance with the input from these components, is *self-legislating*—even if the BCI appears to result in the agent's enhanced instrumental autonomy. My objective is to determine what we may intelligibly conclude about the possibility of integrating a person via a neurotechnological mechanism, as in BCI, for the purpose of enhancing their autonomous agency.

Some have argued that, because various forms of neurotechnology can have a profound effect on the neural basis of the capacities that constitute personal autonomy, the use of such technology precipitates the need to revise, perhaps significantly, the concept of autonomy [1]. Others maintain that to the extent that BCI may address autonomy impairment and given that such technology has as its goal the *restoration* of autonomy that has already been impaired by other factors, there is in fact "no need to revise this concept only to examine the mechanisms behind this restoration" [2]. My view is that this novel form of technology requires at least some consideration of the underpinnings of the notion of autonomy.

Typical cases for the use of BCI are those involving an individual who suffers from severely compromised autonomy due, for example, to a cognitive, physiological or personality disorder, bodily or neurological injury, or some severe form of addiction. In the treatment of such individuals through BCI, brain signals are recorded and used to control some external device that usually takes the form of a computer, prosthetic limb, or semi-autonomous robot [3]. The unique feature of BCI is that unlike other forms of brain-to-computer connection, which require some type of input device, such as peripheral nerves, muscles, a computer keyboard, or mouse, this technology obviates the need for such mediating mechanisms, permitting *direct* brain-computer communication.

In general, three parts constitute a BCI: First, an internal interface enables signal acquisition, second, processing of these signals is carried out through algorithms in a central computing unit, and third, an external interface ensures contact to the surroundings via a specific output device [4].

Mattia and Tamburrini highlight a particularly noteworthy aspect of BCI—*brain-computer mutual adaptation* [5]. One of the components in BCI is the central nervous system; the other is a computer algorithm. Both of these function as *controlling units* and both must be, in some sense, *adaptive*. The type of machine intelligence required in BCI must take the form of "autonomous intelligent action of BCI-controlled devices at least to some degree" [4]. Optimal BCI would also involve substantial BCI-neurofeedback. Jens Clausen et al. note, however, that this type of BCI, which involves bidirectional information flow, has a downside. While optimally effective for integrating brain and computer function, it is also the most problematic for determining the *source* of the choice or act that ensues, which in turn renders it difficult to attribute responsibility for that choice or act [3, 4, 6]. What kind of autonomous agency may be attributed to a being whose capacity to function adequately as an autonomous agent is only rendered possible through a form of BCI technology that appears to inhibit its autonomous agency? The situation is a gripping paradox [7]. Is there any way to break loose?

Neil Levy's work on persons with addictions offers some direction towards a possible resolution. Levy maintains that impaired autonomy is a consequence of the person's fragmented agency [8]. In dealing with this case, Levy appeals to the notion of *extended agency*, which he thinks arises from the possession of a self-governed *will* and a single, unified *self*. Levy's position is very much in keeping with the tradition of thinkers who generally endorse the view that autonomy consists essentially in the agent's capacity to bring their lower-order desires under the control of their higher-order desires. This hierarchical account of autonomy is generally referred to as an *internalist* view. Among its most prominent supporters are Harry Frankfurt [9] and Gerald Dworkin [10]. This tradition is in contrast to the externalist view of autonomy supported by John Christman [11], Alfred Mele [12], John Fischer and Mark Ravizza [13]. Levy's account of *extended agency* has application to the possibility of enhancing autonomous agency through BCI, especially if fully expounded in the context of its historical roots. These roots, I argue, may be located in Immanuel Kant's concept of *autonomy*. The view I advance is that some clarity can be attained on these issues through consideration of the crucial Kantian distinction between *autonomy*, i.e., unified agency, and *heteronomy*, i.e., fragmented agency. I propose that (a) a Kantian-directed account of autonomy facilitates a proper treatment of the possibility of augmenting autonomy through the use of BCI in agential enhancement procedures; (b) this requires an extension of the notion of self-legislation that includes the function of the BCI component among the features of the self at the sub-personal level; and (c) that Kant's concept of the self as an epistemic rather than moral subject may accommodate this extension.

4.2   Instrumental Autonomy and Moral Autonomy

One of the most significant oversights in the literature on debates over neurotechnological and biological enhancement is the conflation of *instrumental* and *moral*
autonomy. The instrumental notion of autonomy is operative in David Hume's [14]
account of human reason, whereas the moral notion of autonomy is at the focus of
Kant's account [15]. Though the two are undoubtedly linked, the difference between
them has important implications for the possibility of augmenting autonomy via
neurotechnology. Instrumental autonomy is the common notion of self-legislation,
indicative of the capacity to function as an *agent*. The BCI that facilitates this function may be described as a type of *agential* enhancement [16]. Agency is typically
construed as the ability to devise and act in accordance with one's own ends, desires
or wishes—regardless of what form they may take. When one is thwarted from
either devising or acting in accordance with those ends, desires or wishes, we tend
to say that the agent's autonomy is compromised or lacking.

In a Kantian framework, however, this kind of agency is not sufficient for the
kind of self-legislation requisite for autonomy. Agency of this kind is referred to by
Kant as *heteronomy*, which is the control of the agent's will by external or alien
forces, i.e., something other than the agent's own *will*. These external forces *include*
the agent's desires, passions, empirical ends, emotions, feelings, and personal
objectives [15]. In Kant's view, the will is only truly self-legislating when not motivated by such external forces. Only then may it be viewed as genuinely autonomous. Kant states that "autonomy of the will is the property of the will by which it
is a law to itself" [15]. He stresses, however, that the *objects* of the will are multifarious in nature; they are mercurial, inconstant, and random. A will guided by external,
i.e., empirical, factors will be similarly inconstant. Moreover, when the objects of
the will conflict, the will too will be conflicted and pulled in a variety of directions.
The result is the fragmentation of the will—a will lacking in unity. It is this fragmentation and lack of unity that characterizes *heteronomy*. In contrast, the capacity
of the will to function as a unified whole is the characteristic feature of *autonomy*.
Only when the agent is motivated in its actions by the cause rather than the effect of
its will is the agent truly self-legislating and therefore autonomous. This happens
only when the will is motivated by the moral law of reason. Construed in this way,
autonomy of the will constitutes for Kant "the supreme principle of morality" [15].

Levy appears to endorse this Kantian conception of autonomy in his claim that,
"autonomy is not freedom or self-realization. Autonomy is self-rule ... An agent is
autonomous to the extent she is able to put her values into effect" [8]. Kant would
add that such values must be determined by reason, which specifies the moral law.
Only the moral law of reason, Kant argues, is a cause that belongs entirely to the
will itself. Kant states:

> If the will seeks the law that is to determine it *anywhere* else than in the fitness of its maxims
> for its own giving of universal law – consequently if, in going beyond itself, it seeks this law
> in a property of any of its objects – *heteronomy* always results [15].

Given that heteronomy consists in the will being motivated by empirical factors, and given that empirical factors amount to natural causes, genuine autonomy, for Kant, must be viewed as the capacity of the will to be affected by a non-natural cause, which originates in the agent's own will. Kant's view of genuine moral agency requires the possession of a will that is able to be governed by a law other than the laws governing the natural world in order to be truly *self-legislating* and not just determined in its actions like all other phenomena in the natural world [15]. This is the law of reason. This non-naturalistic element in Kant's libertarian, incompatibilist moral theory is a real stumbling block for contemporary scholars across the disciplines. Incompatibilism maintains that if determinism is true then free will is impossible. Incompatibilism may take two completely opposing forms: *libertarianism*, which supports free will [17, 18] and *hard determinism*, which denies free will [19]. Kant adopts libertarian incompatibilism since he defends the non-naturalistic causation of reason. A strong majority, however, find this non-naturalistic, rational causation in the Kantian account of autonomy unpalatable. Moreover, it is precisely this non-naturalistic element that would appear to render the enhancement of autonomy through BCI paradoxical.

A related problem to that posed by incompatibilism in BCI use for enhancing autonomy is dualism. Dualism is the metaphysical view that there exist two fundamentally different kinds of substance: material substance (body) and nonmaterial substance (mind). It is generally acknowledged that on a strict dualist view, cognitive enhancement and the enhancement of autonomy would be impossible. Among others, Levy recognizes this problem. He writes, "there is an obvious sense in which cognitive enhancement is a violation of implicit dualism: it produces alterations in the mind via its physical substrate, which ought to be impossible if dualism is true" [20]. Though Kant is not, strictly speaking, a dualist, his position would appear to pose problems for the prospect of enhancing autonomy through BCI similar to those posed by dualism. This is because, for Kant, the *self* has both a *phenomenal* (empirical) aspect and a noumenal (non-empirical) aspect. The noumenal self, though not a distinct *substance* from the phenomenal (empirical) self, is nonetheless not determined by the natural laws that govern the phenomenal self.

Thus, for the above reasons, a Kantian framework does not initially seem promising for augmenting autonomy through BCI. While acknowledging these limitations, I contend that Kant's notion of autonomy can still be rendered serviceable in discussions pertaining to the possibility of enhancing autonomy via neurotechnologies. To be sure, it is not possible to eliminate the non-naturalistic element from the Kantian moral framework without negating Kant's own account of morality altogether. Thomas Hill, however, has argued that Kant's position, though quite distinct from many current accounts of autonomy, nonetheless provides the foundations for some of those accounts [21]. Along the same lines, I maintain that Kantian autonomy can make a significant contribution to debates over the neurotechnological enhancement of autonomy as it incites us to focus on the *cause* of the agent's action rather than the *effect* [22]. Truly autonomous activity, for Kant, is activity motivated

by the right kind of *cause* acting on the will of the agent rather than activity motivated by some object the will happens to desire as an effect. I argue that acknowledging this fundamental feature of autonomy substantially facilitates meaningful
discourse on the possibility of the neurotechnological enhancement of autonomy.
Specifically, it assists in tracking the cause of the agent's will in activity mediated
by brain-computer interfacing. I maintain that the self as epistemic subject in Kant's
theory of cognition serves as the cause of unity required both for cognitive and
moral judgment. Provided, therefore, that autonomy enhancing BCI targets the
cause rather than the *effect* of the agent's action, and that this cause is recognized as
stemming, in part, from the unifying function of the self as *epistemic* rather than
metaphysical subject, it is conceivable how BCI may enhance the *empirical* if not
rational dimension of autonomous agency. Though a thorough treatment of this
issue requires exploring the notion of autonomy in connection with moral responsibility, as noted by Pim Haslaeger [6], the limited scope of this chapter does not
permit a proper consideration of moral responsibility. I focus only on showing how
the Kantian account of self as epistemic subject warrants extending the notion of
self-legislation to render intelligible the enhancement of autonomy through BCI.

Given that instrumental autonomy consists in the capacity for setting and achieving empirical goals, assessing this kind of autonomy involves assessing the agent's
ability to achieve those goals. Thus, if through the use of some BCI an agent is able
to achieve their desired ends more effectively, it would be legitimate to view that
neurotechnological mechanism as having enhanced the agent's *instrumental* autonomy. Assessing the enhancement of moral autonomy is not as straightforward, at
least not if due consideration is given to Kant's grounding moral principle—the
capacity of the will to be motivated by its own internal forces. On this account,
enhancing moral autonomy through BCI would appear to present an insurmountable problem, since it involves the control of the agent's will by an external
mechanism.

Regardless of how efficacious the agent would be as a result of the neurotechnological intervention, the fact that the agent's will is under the partial control of an
external source renders it challenging to view the agent as genuinely *self-legislating*.
According to Kant, agent causation is rational causation and thus, not part of the
series of event causation in nature [23–25]. The question is: Is it possible to conceive of a type of BCI that enhances the will's capacity to be motivated by its own
internal cause, without being affected by an external source? Only BCI that could
operate in this way could come close to meeting the Kantian criteria for autonomy.

The view I defend is that although Kant thinks that *optimal* unity can only be
achieved when the cause of the will is rational rather than empirical, a significant,
though suboptimal, degree of unity may still be achieved when the set of multifarious empirical causes are guided by a higher-order empirical factor capable of bringing them under its control. Despite the fact that this would still be an empirical
rather than rational cause, it nonetheless *functions* in a manner similar in effect to a
rational cause. An empirical cause could conceivably bring this about if it could
subsume the other competing empirical factors under its guidance. To do this, I
maintain, this empirical factor would have to impart the same sort of structural unity

that the self imparts to the multiple features of *cognition* in Kant's theory of knowledge. Thus, a form of BCI that increases unity of will in the agent may be rendered consistent with the general Kantian model of autonomy, though it does not satisfy the non-naturalistic criterion in that model. The non-naturalistic element in Kant's own account of autonomy is ineliminable. Nonetheless, on the view that the general Kantian model really does capture the fundamental feature of autonomy, it would be a significant achievement to be able to approximate maximally that kind of autonomy within the constraints of a naturalistic framework. As I shall show, the workings of this framework may be found in Kant's epistemological program, the focus of which is *the self* as *epistemic subject*.

4.3   Advancing from the Enhancement of Instrumental to Moral Autonomy in a Naturalistic Framework

It seems that virtually all proposals for the enhancement of autonomy are inextricably linked to *cognitive* enhancement [26]. The reasons for this are obvious. Autonomy is typically conceived as requiring a heightened use of reason, and reason is typically construed as a cognitive capacity. The various proposals for the enhancement of autonomy through BCI have in view the strengthening of cognitive capacities in the effort to render reason more dominant than passions and emotions in determining the agent's will. Insofar as reason is guided by the necessary and universal laws of logic, it is considered *more* constant, stable, and fixed [27]. Given that passions and feelings are varied, subjective, and fleeting, a will guided essentially by these empirical factors will inevitably be *less* stable, constant, and fixed. An empirically guided will becomes fragmented—lacking the unity characterized by a will capable of the autonomy required for genuine agency, especially moral agency. Although it has been argued (by non-Kantians) that emotions ought not to be completely eliminated in moral considerations and that they in fact play a fundamental role in any morality fit for human beings, it does not follow that it is inappropriate for reason to regulate emotions. Reason may effectively guide the will in deliberating on the degree to which certain emotions ought to influence the will in a given context. Moreover, some claim that Kant's focus on the rational capacity of persons generates a very narrow conception of personhood that neglects its relational aspects [28]. I maintain that, even acknowledging these relational aspects, and the importance of the role of emotions, reason's status is paramount. Reason functions as the force that unifies the will as it governs these relational and emotional aspects of the person; and this function can be facilitated by reliable cognitive faculties. Thus, neurotechnological programs directed at enhancing autonomy by reinforcing the agent's cognitive capacities are certainly on a promising path.

I want to argue, however, that enhancing cognitive capacities is not *sufficient* in getting us beyond mere *instrumental* autonomy to a substantive form of *moral* autonomy. The reason is that in the instrumental function of reason there is considerable potential for heteronomy. That is, instrumental reason is hypothetical in nature. It operates on the precept "If x, then y." That is to say, "if one wants to achieve a particular desired end, x, then one ought to do y." Kant refers to this kind of principle as a "hypothetical imperative." In contrast, he maintains that morality is characterized by a "categorical imperative." A categorical imperative takes the form "Do y, necessarily." It commands without a view to some desired end and is therefore not prone to the instability that results from having one's actions determined by such ends [22]. Only a categorical imperative, Kant asserts, can serve as a genuine moral principle that may govern the will necessarily, and thus only such an imperative may serve to unify the will in the manner required for moral autonomy [15]. To the extent that cognitive enhancement only aims at fostering instrumental autonomy, i.e., hypothetical reasoning, I maintain that cognitive enhancement is not a direct or guaranteed route to moral enhancement [29]. I provide a more detailed defense of this position elsewhere [30].

Even if cognitive enhancement could be effectively achieved through BCI, a fundamental problem still remains. As Jens Clausen et al. state,

> although effortless interactions between mind and machine seem intuitively appealing, creating direct links between a digital machine and our brain may dangerously limit or suspend our capacity to control the interaction between the 'inner' personal and outer worlds [31].

The interaction between a human agent's brain and an intelligent machine renders it difficult to determine the *source* of an act [31]. Moral autonomy requires that this source be the agent's will and absolutely nothing else. The challenge, therefore, in the use of BCI to augment moral autonomy consists in somehow integrating the deliberation of the intelligent machine into the deliberation of the agent's will, such that the two are unified into one being and thus one source. Unless this can be brought about, the machine component in BCI will remain *other* and as such could only serve to subvert the agent's autonomy rather than augment it. In the section that follows, I propose a strategy to address this problem.

4.4 Escaping the Paradox: Extending the Notion of Self-Legislation

The paradox that confronts us is: How can it be possible to enhance an agent's moral autonomy by means of a factor external to the agent's will if the very nature of moral autonomy requires that the agent *not* be affected by external factors. Lucivero and Tamburrini have also drawn attention to this paradox in BCI:

> There are various ways in which the inclusion of a robotic controller, say, in the motor pathway of an output BMI [Brain Machine Interface] may limit or jeopardize personal autonomy. It is quite paradoxical that these threats to personal autonomy may arise in systems, which are mostly designed and implemented for the purpose of protecting and promoting personal autonomy by restoring lost motor functions and re-establishing the capability of interacting with the external environment [7].

What Lucivero and Tamburrini refer to as *personal autonomy* I refer to as *instrumental autonomy*. While they raise the paradox in the context of autonomy in general, I maintain that the paradox is particularly problematic for moral autonomy. The only route of escape, I argue, is to adopt an *extended* notion of *self-legislation*.

In order to fulfill the requirements for self-legislation, the robotic controller to which the agent's brain is coupled must be viewed as an *extension* of the agent's will (on a compatibilist view) and therefore, as an extension of its autonomous function through self-legislation. Given the agent's *dependence* on the computer component to generate the conditions for the unity that moral autonomy requires, this extension is warranted. Only by acknowledging the validity of the agent's dependence on the BCI components for producing the requisite conditions for self-legislation is it possible to avoid the paradoxical conclusion that the agent's will is being determined by an external source. The resolution of the paradox thus requires the bold tenet that self-legislation can be extended to include the function of an artificial intelligent being via BCI. The agent's autonomy is enhanced as their will becomes integrated in the *closed-loop system* generated as the agent incorporates the feedback from the BCI in their deliberation. In consequence, the agent achieves a higher level of control. Michael Bratman's account of *temporally* extended agency bears some resemblance to the core features of my argument here [32]. It differs from my account, however, in focusing on agency extending over time intervals rather than shared control systems. Lucivero and Tamburrini illustrate the operation of such shared control systems in the following:

> The subject issues high-level control inputs for the robotic controller, which the brain reading components of this BMI [Brain Mind Interface] extract from EEG signals produced through the voluntary execution and control of some mental task. Low-level commands, concerning the detailed trajectory of the controlled robotic device are issued independently by the robotic controller. In addition to this, one should be careful to note that in this output BMI the higher-level control of robotic action is shared too, insofar as it results from the combined processing of EEG data, robotic sensor, and memory traces [7].

The key factor in such *shared control systems* is the distinction between low-level commands and high-level commands. I maintain that as long as the robotic controller only initially contributes low-level commands and does not contribute to the high-level commands until after some significant integration with the person's own high-level commands, we are warranted in viewing this activity as self-legislation on the part of the agent.

Needless to say, the ramifications of this position are quite momentous as it appears to entail the requirement of a radical rethinking of our concept of *self.* My concerns here are not rooted in the general program pursued by John Locke [33], David Hume [14], Sidney Shoemaker [34], Derek Parfit [35], and more laterally explored in connection with BCI by Federica Lucivero and Guglielmo Tamburrini [7]. Their programs investigate issues pertaining to the continuity of personal identity over time and throughout change, i.e., diachronic identity and to some extent synchronic identity, i.e., identity at a time. I limit myself here only to the challenge of tracking a person's *self* through the capacity for self-legislation in individuals participating in BCI. Tamburrini seems to suggest that what this essentially requires is the acknowledgment of the distinction between the *personal* and *sub-personal* self. He maintains that on some interaction models of brain-to-computer communication, the relevant communication occurs on some *sub-personal* level. Tamburrini describes this level as one that does not involve "the intentions, beliefs, and contents of consciousness" [36]. As Tamburrini explains,

> cooperative human-computer problem solving systems of this kind pave the way to what one may appropriately call a subpersonal use of human beings in HCI [Human Computer Interaction], insofar as the role of human beings involved in this cooperative task is accounted without appealing to any of the more characteristic features of human mentality and personhood [36].

Daniel Dennett was among the first to suggest the distinction between the personal and sub-personal levels of the self [37]. Though Dennett later came to reject this distinction, Jennifer Hornsby argues that he ought to have retained it, given its utility [38].

The importance of the distinction between the personal and the sub-personal also appears to be operative in Levy's solution to the problem of addiction. Here the control of the personal self over the various sub-personal components constitutes extended agency. Levy suggests that the notion of *extended agency*, which involves the related notion of the *extended will*, may be appealed to in the attempt to render intelligible the Aristotelian notion of *akrasia*, i.e., weakness of will or incontinence [39]. *Akrasia* asserts itself in cases of what Levy calls "unwilling addiction," i.e., cases in which "the addict acts against her will, even though she chooses consumption, and values it when she chooses it" [8]. The account Levy provides of extended agency preserves some of the most essential features of the Kantian account of autonomy: will and personhood. Levy suggests that we

> identify the agent's will with her *extended agency*. Being an agent, that is, having a single, relatively unified self, is not something to which we are simply born. Instead, it is an achievement. We gradually unify ourselves [8].

This view is also defended by Christine Korsgaard in her Kantian account of *self-constitution* [40]. Levy acknowledges the heteronomy that plagues the agent's attempt to cultivate a genuine will and that promotes instead the fragmentation of the will. He asserts,

> our minds are built up out of a large number of sub-personal mechanisms, which differ in the extent to which they receive input from each other and from consciousness [8].

Implicit in Levy's account is the influence of the heteronomous factors to which Kant refers that threaten the unity of the will, and thereby the unity of the self. These involve spurious principles rooted in the objects of desire, emotion, and passion that compete with reason in governing and motivating the will. This incapacity of an agent to act in accordance with their own genuine will, motivated instead by sub-personal factors over which they lack control and thus do not recognize as deriving from their own true identity is precisely what is targeted by BCI. It should be noted, however, that it is often expected that the *effects* of BCI will be manifested on the personal level, especially in the treatment of psychiatric disorders [41].

Given these considerations, I argue that to resolve the paradox of enhancing autonomy through BCI, either we must revise the notion of autonomy to mean something other than self-legislation or we must revise the notion of self-legislation. I suggest we retain the notion of autonomy as self-legislation, since it is the only meaningful one, and that we extend the notion of self-legislation to include the components and functions of BCI. My general argument here is analogous in some respects to David Chalmers and Andy Clark's argument in defense of the *extended mind*, but I do not think that many of the objections that have been presented against their position pose quite the same problems for mine [42].

A crucial qualification of my position is that this extension should not occur at the level of self as personal, but only at the level of self as sub-personal. Specifically, it is reasonable to extend the sub-personal level of the self to include the technological mechanisms to which the self is coupled via BCI as additional parts of the sub-personal level of the self. Again, however, we should expect that the effects may be *manifested* at the personal level, as the enhancement may result in some changes in personality or lifestyle. This idea should not be any more extraordinary than extending the sub-personal level to include components that are not part of a person's original self in cases such as an organ transplant, a blood transfusion, or the acquisition of a pacemaker or prosthetic limb. As long as it is possible to track the agent's will on a personal level, there is no reason to think the robotic controller to which the person is neurotechnologically coupled could not legitimately be considered a constituent part of the agent on a sub-personal level. Given the strong interaction between them and the feedback loop this generates, there is a substantial and reasonable basis for extending the sub-personal self in this way. Further justification for this is the claim defended by Wolkenstein, Jox, and Friedrich that "in BCI use, the user (i.e., the one who uses an algorithm to make decisions), and the entity mostly dealing with the effects of the algorithm, is one and the same person" [43]. Moreover, Levy contends that "agents have inclination, and inclination must have neural realizers" [20]. Even Kant seems to suggest a connection between the two conceptions of the self that feature in his account. Kant distinguishes sharply between the *noumenal* (metaphysical) self, i.e., a being possessing intelligence and will, and the *phenomenal* (physical) self, i.e., the body [44]. He maintains however that there is some kind of association between them [22, 44]. What Kant would rule out is the possibility of direct causal action of the robotic components on the noumenal self as though the two were each links in the chain of natural causation. This has led some scholars to think that Kant endorses a moderate metaphysical view of the self. I contend that this view accommodates an open-ended self that is able to be expanded to include the right kinds of components beyond the person's own biological body, provided those components contribute to the unity of the self.

Not surprisingly, consensus is lacking among Kant scholars on how exactly Kant's notion of the self is to be understood—especially as it is used in many senses, depending on the various contexts. Béatrice Longuenesse provides a detailed account of the various senses of Kant's concept of the self [45]. Regardless of which of these senses is considered, Kant's crucial claim about the self is that, as in the case of all objects of our knowledge, we do not have an understanding of the self as it really is *in itself* [44]. Kant develops this position in his critique of his rationalist predecessors, most notably René Descartes, who maintained that we *do* have knowledge of the self as it is in itself—a single, simple, enduring, metaphysical substance [46]. In contrast to Descartes, Kant argues that we do not possess any fundamental knowledge of the *real essence* of the self. Lacking this knowledge, I maintain, we also lack a basis for rejecting offhand the view that the self could be extended to include the neurotechnological devices and robotic components involved in BCI.

Although in his ethical program Kant maintains that morality requires us to conceive of the self as a metaphysical being capable of rational causation [22], in his epistemological program he stresses that we may not have any knowledge of the nature of this rational being, only of its function in cognition [44]. As Karl Ameriks notes, Kant emphasizes "our epistemic essence as rational epistemic subjects capable of determining [cognitive] truths" [47]. In this sense, the self has capacities that take the form of structures for determining the fundamental features of our knowledge. This sense of self captures not some non-natural, metaphysical essence, but rather our built-in system of structures for cognizing the world. According to Kant, "these structures come from the 'self' or are 'due to us' in that they are not simply the result of anything understood solely as an outer – merely supernatural or natural – force" [47]. The key feature of these structures is that they are responsible for the *unity* of our judgments, and ultimately the unity of our knowledge. It is in being the source of epistemic unity that the self is *itself* a unity [22]. Similarly, the components of BCI designed to enhance an agent's autonomy may legitimately be considered part of the self on the basis that they too function as structures that contribute to the unity of the self as an epistemic subject, in so far as they enhance the cognitive function of the agent, thereby making them more capable of and responsive to reasoning, i.e., more self-legislating. As Wolkenstein, Jox, and Orsolya stress, the algorithm in BCI "directly connects with the intention-generating and, therefore, action-initiating processes within a person's brain. A BCI is constitutive of an action and not merely a replaceable instrument" [43]. It is for this reason that the relation between person and artificial component in BCI is one of *integration* and not merely *interaction*. This is rendered significantly more plausible in cases where the integration through BCI involves a reliable feedback loop, which fosters adaptiveness on the part of the artificial components. It is in this way that the Kantian notion of *self* as *epistemic subject* serves as a helpful model for understanding how the notion of *self* could be extended to include the sub-personal components in BCI, and thereby ground the extension of *self-legislation* required to render intelligible the enhancement of autonomy through BCI.

Undoubtedly, one of the most serious challenges this position confronts is the demarcation problem. Once we start extending the boundaries of the self to include external, artificial components, how do we determine what is to be included within the boundaries in a legitimate and non-arbitrary fashion [48]? Fred Adams and Ken Aizawa [49] pose an objection to the extended mind thesis that may appear to have some application to my argument for extended self-legislation. Defending an internalist theory of cognition, Adams and Aizawa argue that only brain states have a natural kind of content that is intrinsic to the brain states themselves, rooted in personal history and not legitimately attributable to external resources that are merely used as cognitive tools. The reason these types of arguments are not as effective in the case of extended *self-legislation* as in *extended mind* is that on a Kantian view of the cognitive subject there is no determinate, biological or metaphysical basis of this subject. The only fundamental requirement is cognitive unity, and if that can be enhanced through BCI then the components of BCI are legitimately within the bounds of the extended self, while all other external factors remain safely beyond those bounds. Moreover, David Kaplan [50] presents a response to Fred Adams and Ken Aizawa's objections to the extended mind that I think is also applicable here. As Kaplan maintains: "one might … define the boundaries of cognition in terms of the pattern of causal interactions among the parts within a given system" [50]. The degree of causal connectedness and interaction between the parts of the system, natural or artificial, is, I think, the determining factor for inclusion in the *self-system*. Moreover, this connectedness and interaction is most effectively achieved between the natural and artificial components at the sub-personal level of the self.

Some, however, maintain that neurotechnological intervention may take a more radical form. Alberto Giubilini and Julian Savulescu advocate the development of an AI mechanism that functions as an "artificial moral advisor" [51]. This mechanism would be a highly intelligent artificial being designed to provide expert advice in an extremely efficient manner to facilitate both an agent's deliberation and action on moral matters. In addition, this artificial moral advisor would guide the agent in determining which emotions ought to come into play and which ought to be suppressed in a particular instance. Giubilini and Savulescu stipulate, however, that this artificial moral advisor ought not to impose any particular moral theory on the agent, rather it would employ the agent's own moral principles to carry out the moral deliberation. Initially, this would certainly appear to step beyond the appropriate boundaries of even an extended notion of self-legislation. It should be noted, however, that the general conception of a *moral advisor* has considerable affinity with Kant's notion of the *innate judge* that functions as a tribunal or court and that dwells in every person. Kant describes this feature of a person as the "consciousness of an *internal court* in the human being ('before which his thoughts accuse or excuse one another')" [52]. What Kant is referring to here is *conscience*. The proper administration of this court, Kant explains, requires that a person conceive of "*someone other than* himself (i.e., other than the human being as such) as the judge of his actions, if conscience is not to be in contradiction with itself. This other may be an actual person or a merely ideal person that reason creates for itself" [52].

The question that confronts us is: could this feature of *persons* be substituted, or at least augmented, by an artificial intelligent being? At first consideration, it would seem not, given that the artificial being is *external* to the agent. If, however, this artificial being were to be *coupled* with the person in such a way that it could be conceived as a constituent *part* of the person at the sub-personal level, there may be a basis for viewing this artificial component as internal rather than external to the agent and therefore, as not posing an obvious threat to the person's autonomy. Having been integrated with the person in this fundamental way, the artificial moral advisor would augment the function of the moral law of reason that serves to unify the will of the agent, which in turn renders the agent more morally autonomous. In this way, it might be argued, the artificial moral advisor facilitates the move from merely instrumental to moral autonomy. Facilitating this move, however, is not equivalent to *guaranteeing* it. There may be a variety of other conditions that need to be met before this move is efficacious. With the right set of limiting conditions however, perhaps even something as extreme as a BCI that functions as a *moral advisor* is not completely out of the question for programs aimed at enhancing the autonomy of persons. Nonetheless, whether the requisite degree of integration between the self and any type of artificial component in any type of BCI can *actually* be achieved remains, as yet, uncertain. My aim has only been to determine whether it is at least *conceptually* possible.

4.5   Conclusion

I have argued that on an extended view of *self-legislation,* based on a Kantian conception of the self as epistemic subject, there is legitimate basis for considering the components of BCI as an extension of the self on a sub-personal level. The self as epistemic subject, however, must be construed as operating on a fundamental cognitive level, not a merely instrumental level. It must be considered the active source of the structures that contribute unity to cognition, not merely an instrument of prudential judgment. Lacking any knowledge of the essence of the epistemic self, except that it is responsible for the required unity of our knowledge, there is no reason for immediately rejecting the view that the components of BCI could be incorporated in this notion of the self. The requisite integration, however, only occurs on a sub-personal level of the self, and can thus, at most, enhance the cognitive capacities necessary for generating the unity of will that grounds moral autonomy. It is in this qualified sense that the enhancement of moral autonomy through BCI may be rendered intelligible. Whether BCI can actually be designed to achieve this end will be among the most anticipated answers that researchers in neurotechnology and artificial intelligence will be tasked to provide.

References

1. Müller S, Walter H. Reviewing autonomy: implications of the neurosciences and the free will debate for the principle of respect for the patient's autonomy. Camb Q Healthc Ethics. 2010;19:205–17.
2. Glannon G. Neuromodulation, agency and autonomy. Brain Topogr. 2014;27:46–54.
3. Clausen J. Bonding brains to machines: ethical implications of electroceuticals for the human brain. Neuroethics. 2013;6:429–34.
4. Clausen J. Ethical implications of brain-computer interfacing. In: Clausen J, Levy N, editors. Handbook of neuroethics. Dordrecht: Springer; 2015. p. 699–704.
5. Mattia D, Tamburrini G. Ethical issues in brain-computer interface research and systems for motor control. In: Clausen J, Levy N, editors. Handbook of neuroethics. Dordrecht: Springer; 2015. p. 725–40.
6. Haselager P, Did I do that? Brain-computer interfacing and the sense of agency. Mind Mach. 2013;23:405–18.
7. Lucivero F, Tamburrini G. Ethical monitoring of brain-machine interfaces a note on personal identity and autonomy. AI Soc. 2008;22:449–60.
8. Levy N. Autonomy & addiction. Can J Philos. 2006;36:427–47.
9. Frankfurt H. The importance of what we care about. Cambridge: Cambridge University Press; 1988.
10. Dworkin G. The concept of autonomy. Grazer philosophische studien. 1981;12/13:203–13.
11. Christman J. Autonomy and personal history. Can J Philos. 1999;21:1–24.
12. Mele A. Autonomous agents: from self-control to autonomy. Oxford: Oxford University Press; 1995.
13. Fischer J, Ravizza M. Responsibility and control: a theory of moral responsibility. Cambridge: Cambridge University Press; 1998.
14. Hume D. In: Selby-Bigge LA, editor. A treatise of human nature. Oxford: Oxford Clarendon Press; 1896.
15. Kant I. Groundwork of the metaphysics of morals. In: Guyer P, Wood AW, editors. The Cambridge edition of the works of Immanuel Kant. Cambridge: Cambridge University Press; 1992.
16. Earp BD, Douglas T, Savulescu J. Moral neuroenhancement. In: Johnson LSM, Rommelfanger KS, editors. The Routledge handbook of neuroethics. New York: Routledge; 2017. p. 166–84.
17. Van Inwagen P. An essay on free will. Oxford: Oxford University Press; 1983.
18. Kane R. Free will and values. Albany: State University of New York Press; 1985.
19. Pereboom D. Living without free will. Cambridge: Cambridge University Press; 2001.
20. Levy N. Neuroethics: ethics and the sciences of the mind. Philos Compass. 2009;4:69–81.
21. Hill T. Kantian autonomy and contemporary ideas of autonomy. In: Sensen O, editor. Kant on moral autonomy. Cambridge: Cambridge University Press; 2013. p. 13–21.
22. Kant I. Critique of practical reason. In: Guyer P, Wood AW, editors. The Cambridge edition of the works of Immanuel Kant. Cambridge: Cambridge University Press; 1992.
23. Chisholm R. Freedom and action. In: Lehrer K, editor. Freedom and determinism. New York: Random House; 1970. p. 11–40.
24. Taylor R. Action and purpose. Englewood Cliffs: Prentice Hall Publishing; 1966.
25. O'Connor T. Agent causation. In: O'Connor T, editor. Agents, causes, and events: essays on indeterminism and free will. New York: Oxford University Press; 1995.
26. Schaefer GO, Kahane G, Savulescu J. Autonomy and enhancement. Neuroethics. 2014;7:123–36.
27. Shook JR. Neuroethics and the possible types of moral enhancement. AJOB Neurosci. 2012;3(4):3–14.
28. Burwell S, Sample M, Racine E. Ethical aspects of brain computer interfaces: a scoping review. BMC Med Ethics. 2017;18:1–11.
29. Schaefer GO. Direct vs indirect moral enhancement. Kennedy Inst Ethics J. 2015;25:261–89.

30. Wilks AF. Kantian challenges for the bioenhancement of moral autonomy. In: Hausekeller M, Coyne L, editors. Moral enhancement: critical perspectives, Royal Institute of Philosophy Supplement 83. Cambridge: Cambridge University Press; 2018. p. 121–43.
31. Clausen J, Fetz E, Donoghue J, Ushiba J, Spörhase U, Chandler J, Niels Birbaumer and Surjo R. Soakader. Help, hope, and hype: ethical dimensions of neuroprosthetics. Science [Internet]. 2017;356:1338–9. http://science.sciencemag.org/content/356/6345/1338.
32. Bratman M. Reflection, planning, and temporally extended agency. Philos Rev. 2000;109:35–61.
33. Nidditch PH, editor. John Locke: An essay concerning human understanding. Oxford: Oxford Clarendon Press; 1975.
34. Shoemaker S. Personal identity. Hoboken: Blackwell; 1984.
35. Parfit D. Reasons and persons. Oxford: Oxford University Press; 1986.
36. Tamburrini G. Brain to computer communication: ethical perspectives on interaction models. Neuroethics. 2009;2:137–49.
37. Dennett D. Content and consciousness. London: Routledge; 1969.
38. Hornsby J. Personal and sub-personal: a defense of Dennett's early distinction. Philos Explor. 2000;3(1):6–24.
39. Aristotle. Nicomachean ethics. In: Barnes J, editor. The complete works of Aristotle. The revised Oxford translation, vol. 2. Princeton: Princeton University Press; 1984. p. 1808–25.
40. Korsgaard C. Self-constitution: agency, identity, and integrity. Oxford: Oxford University Press; 2009.
41. Clausen J. Conceptual and ethical issues with brain-hardware interfaces. Curr Opin Psychiatry. 2011;24:495–501.
42. Chalmers D, Clark A. The extended mind. Analysis. 1998;58:7–19.
43. Wolkenstein A, Jox RJ, Friedrich O. Brain-computer interfaces: lessons to be learned from the ethics of algorithms. Camb Q Healthc Ethics. 2018;27:635–46.
44. Kant I. Critique of pure reason. In: Guyer P, Wood AW, editors. The Cambridge edition of the works of Immanuel Kant. Cambridge: Cambridge University Press; 1992.
45. Longuenesse B. I, me, mine: back to Kant, and back again. Oxford: Oxford University Press; 2017.
46. Descartes R. Meditations on first philosophy. Cress DA, translator, editor. Hackett: Indianapolis; 1993.
47. Ameriks K. Kant and the fate of autonomy. Cambridge: Cambridge University Press; 2000.
48. Levy N. Rethinking neuroethics in the light of the extended mind thesis. Am J Bioeth. 2007;7:3–11.
49. Adams F, Aizawa K. Defending the bounds of cognition. In: Menary R, editor. The extended mind. Cambridge: MIT press; 2010. p. 67–80.
50. Kaplan DM. How to demarcate the boundaries of cognition. Biol Philos. 2012;27:545–70.
51. Giubilini A, Savulescu J. The artificial moral advisor: the "ideal observer" meets artificial intelligence. Philos Technol. 2018;31:169–88.
52. Kant I. Metaphysics of morals. In: Guyer P, Wood AW, editors. The Cambridge edition of the works of Immanuel Kant. Cambridge: Cambridge University Press; 1992.

Can BCIs Enlighten the Concept of Agency? A Plea for an Experimental Philosophy of Neurotechnology

5

Pim Haselager, Giulio Mecacci, and Andreas Wolkenstein

Contents

Abstract

Passive BCIs can be used to measure brain processes that take place without necessarily having the intention to communicate, or even while being unaware that specific information about mental states is being collected. This type of symbiotic neurotechnology has the potential to create new and philosophically fascinating cases where the question of "was that me?" will make sense from both an

P. Haselager (✉)
Donders Institute for Brain, Cognition, and Behaviour, Department of Artificial Intelligence, Radboud University, Nijmegen, The Netherlands
e-mail: w.haselager@donders.ru.nl

G. Mecacci
Department of Artificial Intelligence, Radboud University, Nijmegen, The Netherlands

Department of Values, Technology and Innovation, Faculty of Technology, Policy and Management, TU Delft, Delft, The Netherlands
e-mail: g.mecacci@donders.ru.nl

A. Wolkenstein
Institute of Ethics, History and Theory of Medicine, Ludwig-Maximilians-Universität (LMU) München, Munich, Germany
e-mail: andreas.wolkenstein@med.uni-muenchen.de

© Springer Nature Switzerland AG 2021
O. Friedrich et al. (eds.), *Clinical Neurotechnology meets Artificial Intelligence*, Advances in Neuroethics, https://doi.org/10.1007/978-3-030-64590-8_5

individual and a societal perspective. We think that symbiotic technology is philosophically interesting in that it enables subconscious brain states to influence actions in a new, technology-mediated way. We will examine some of these cases and make a plea for a more systematic use of symbiotic technology in experimentally guided thought experiments aimed at studying the sense of agency. Our guiding questions are: What could technology-induced agency confusions tell us about the experience of ownership of action? What theoretical (e.g., conceptual) and practical implications (e.g., related to identity and responsibility) might this have?

5.1   Introduction

The distinction between something that I do and something that happens to me is generally clear, but not always an easy one to make. Increasingly we are embedded in environments full of artificial "helpers" that support our thoughts and actions in various ways, e.g., by sending us reminders or by proactively presenting us with information that we might consider relevant. Such artificial support systems can be brain-based. Neurotechnology creates the possibility of deriving information about mental states and preferences from brain measurements. In the case of active BCIs, brain processes are related to a specific intention that is meant to communicate information, which occurs, for instance, in cases of deliberate mental imagery, e.g., to write a text or move a robot. Passive BCIs can be used to measure brain processes that take place without necessarily having the intention to communicate, or even while being unaware that specific information about mental states is being collected. This type of symbiotic neurotechnology has the potential to create new and philosophically fascinating cases where the question of "was that me?" will make sense from both an individual and a societal perspective. We will examine some of these cases. We make a plea for the use of symbiotic technology in experimentally guided thought experiments that aim to clarify and refine our understanding of the sense of agency.

5.2   Neuroadaptation and Symbiotic Technology

Efficient technology is designed to get in the way as little as possible. Therefore, passive BCIs have great potential in that the brain activity that they measure provides implicit input for a computer, and thus does not require any effort to communicate [1–3]. Passive BCI implementations use natural brain activity to infer aspects of a user's cognitive or affective state, without needing the user to consciously attempt to communicate such information to the system. They are the type of BCIs where users can "loosen the reins" [4, 5] and allow their brain states to be used to enable applications that achieve better performance, require less user effort, and result in less interference with ongoing cognition. When using this information to create a model of the user that can form the basis for user-supporting actions, one can speak of neuroadaptive or symbiotic technology (ST). The phrase "neuroadaptive" emphasizes that the system adjusts itself on the basis of the user's brain activity. The label "symbiotic technology" may be taken to stress the functional integration of both user and system in task performance [6]. As we are focusing on the implications of human actions mediated by intelligent technology, rather than the technical aspects of how such systems can adapt to their users, we will use the label "ST" in the remainder of this chapter.

We suggest that STs can be seen as utilizing indirect, implicit, probed, and extrapolated information about a user's cognitive processes and states. The information is *indirect* because it is based on brain measurements. As there are usually many choices, assumptions and inferential steps that need to be made to interpret brain measurements—steps that make use of algorithms and are thus liable to how well the algorithm works—, deriving this information implies traversing a huge inferential distance [7–9]. Second, the information is usually *implicitly* communicated, in the sense that users do not know or are unable to control what data gets acquired [10]. ST enables

> an intimate kind of relation that is made possible by the synergic advances in physiological computing, biometrics, sensing technologies, and machine learning, often combined with the ubiquity of networked devices. Users do not necessarily need to be aware of what is happening while machines help themselves with information [11].

This can bypass, for instance, a user's possibility to direct or veto the process [12, 13]. Third, STs can actively sample information by generating a *probe*, i.e., providing input that induces a brain response that aims to acquire specific information about the user [2]. Finally, the information is *extrapolated* in the sense that the ST normally builds and continually updates a model of the user that forms the basis for the system's contribution to human action.

STs can be used in many different ways and for many different purposes. Just to mention a few possibilities, Thorsten Zander and colleagues [14] indicate that EEG-based passive BCIs (pBCIs) have been used to infer a specific user intention [15, 16], situational interpretation [17], or a change in cognitive [18] or affective [19] states, in real time. STs can collect information in order to *characterize* the user and to generate a diagnosis about cognitive features, e.g., preferences. It can provide a particular type of *feedback* about brain processes to the user, as in neurofeedback. It can support a user in *pragmatic action* [20], as in traditional BCIs, i.e., in order to help users achieve a desired change in their physical environment. Finally, it can assist a user in *epistemic action* [20], which is when users create new perceptual input for themselves in order to achieve a cognitive change, usually related to the exploration of a particular problem-solving task (e.g., finding solutions by meaninglessly manipulating blocks in the computer game Tetris).

5.3   Sense of Agency

If one's cognitive processes and actions are mediated (e.g., supported, nudged, assisted) based on information derived from brain processes outside the control and awareness of oneself, how would that affect one's sense of agency and ownership of action? To what extent could such technology affect self-determination, and what consequences would arise for responsibility and accountability?

The experience of doing something, such as being engaged in an action, can be contrasted with the experience of something happening to oneself. The former, which is referred to as a sense of agency (SA), has been described by Shaun Gallagher [21] as: "The sense that I am the one who is causing or generating an action." This can be distinguished from sensing that one is being moved or effectuated by other factors, e.g., being pushed. Reflexes, i.e., movements performed involuntarily in response to external stimulation such as a knee-jerk or a blink, form an interesting in-between category.

Upon closer analysis, several different aspects of the sense of agency can be distinguished. For instance, there is a difference between a sense of movement (SoM: ± "I'm moving" vs. "something is moving around me") and a sense of generating movement (SoGM: ± "I'm generating my movement" vs. "something is setting me in motion," e.g., the difference between stepping to the side vs. being pushed to the side). And these can be distinguished again from a sense of wanting to move or from trying to move. We will be focusing on the distinction between a relatively low-level implicit feeling of agency (FoA), a non-conceptual, immediate, ongoing "dim" appreciation of acting, and a judgment of agency (JoA) consisting of a relatively high-level, conceptual, reportable, possibly retrospective inference about being the author of an act [22–24].

Under some circumstances, a sense of agency can be incorrect, as Daniel Wegner has suggested in his seminal "The illusion of conscious will" [25]. Wegner describes the experience of conscious will as potentially erroneous regarding the causal efficacy of the will, and suggests that it should be seen as "a compass" (p. 341) for our claims to have engaged in an action. Conscious will is an "authorship emotion," a "somatic marker" of personal authorship of an action. Its embodied quality gives the experience of will a more profound quality than a "mere" thought. It can intensify our appreciation of what we are doing and our memories of what we have done. Moreover, it is a guide to our moral responsibility for our own actions and their consequences ([26], p. 325). It shapes our sense of identity in that we learn about ourselves through experiencing our actions. Normally, Wegner suggests, our awareness of actions and our actions are in synchrony. We feel that we are doing something when we act, and we don't when we don't. However, under special circumstances this correlation may become undone. In cases of automatism, e.g., delusions of alien control, we might actually be the initiators of actions, yet not experience them, leading to the attribution of our actions to outside agents. For instance, it has been suggested that patients with schizophrenia that report hearing voices actually produced the vocalizations themselves without actually experiencing doing so [27]. In the context of BCI, you could causally contribute to the production of an output, e.g., in virtue of your brain's P300 signal, without having the feeling of doing anything.

Alternatively, there can be situations where we have the experience of doing something, while we are in fact not part of the causal chain of events. A simple example here might be the experience of putting something down on a table or dropping something on the floor, while at exactly the same moment a loud noise can be heard. For a brief moment, we can feel as though we are responsible for the loud noise, as evidenced for instance by briefly cringing, while in reality we were not responsible. Another example would be going through the motions of opening a door, while at the same time someone on the other side is opening it for you. You may not have opened the door in the sense of exerting enough force for it to open, but until you find out who actually opened the door you might feel like you were the one who opened it.

Despite its relative straightforwardness and experiential simplicity, the sense of agency appears to rely on a rather complex constellation of underlying processes. The sketch we provide here is based on the paper by Matthis Synofzik and colleagues [28] and provides a rough outline of the various contributing factors. First of all, one can identify a "feeling of agency" (FoA), which is a pre-reflective, non-conceptual and often implicit sense of being engaged in action. Secondly, there is a reflective, conceptual "judgment of agency" (JoA) that consists of a more explicit awareness of acting, contributing to a narrative that consists of reasons for acting [21, 29, 30]. The comparator model by Frith [21, 31, 32] can be considered to form the basis for the FoA. Its essence consists of a comparison between the predicted sensory consequences of an action (the efferent copy computed by a forward sensory predictor system) and the actual sensory feedback. The match between sensory prediction and the actual sensory input generates the FoA. Wegner and Wheatley [33] suggested a more high-level inferential process underlying the JoA. When an intention to act precedes (within a reasonable time frame) the bodily movement (priority), and the movement is compatible with the thought (consistency), and there are no other plausible explanations for the movement (exclusivity), then the movement is inferred to be one's own action. Synofzik et al. [24, 28] suggest that both systems are integrated adaptively, depending on contextual cues, and so the overall sense of agency depends on a complex and sophisticated predictive and postdictive interplay between various brain processes influenced by contextual cues. The sense of agency therefore is not to be considered as one relatively straightforward system. At least two aspects of it can be distinguished (FoA and JoA) and both are thought to involve complex processes that are interconnected, constantly balancing several different authorship cues according to their relative reliability in a given situation.

In two earlier papers [34, 35] one of the authors of this chapter examined the possible consequences of active BCIs for a user's judgment of agency and briefly reported a pilot study with a setup that is comparable to Wegner's Helping Hands experiments [36]. In a standard BCI setup, participants were asked to imagine a hand movement while watching the effects of their mental imagination on a robotic hand. The preliminary results suggested that, at least under some conditions, users could report a degree of agency even in those cases where their brain states played no role in the causal chain leading to the action. One of the participants even offered an excuse for failure to perform the task correctly by claiming that they were distracted [37]. The two papers speculatively considered the consequences of adding an additional confusing factor to the BCI, i.e., an intelligent device (ID) such as a wheelchair with obstacle avoidance or path-planning capacities. Given that the overall behavior would depend on the combined activity of three different types of intelligences (human, BCI, ID), more agency confusions could be expected and ultimately responsibility for action might become diffused [38]. Those papers did not discuss the consequences of passive BCI-based symbiotic technology, but focused on situations where users were actively engaging in mental activity (mental imagery), which intended to communicate instructions to the BCI via their brain states. Here, we would like to offer some thoughts on the consequences of the usage of passive BCIs that do not rely on explicit attempts to communicate, but that use brain signals that do not correlate with an awareness of an intention to do something. Our aim is to investigate whether experimentally induced confusions of the SoA could tell us about the processes involved. In addition, what could the consequences be for the user's perception of actions mediated by ST in terms of agency? Would they regard them as not being their own? What kind of appropriation, what stamp of approval would or could they give them? Would such actions be experienced as being comparable to a reflex (like a blink), an intended action (a wink), or something completely unrelated (an event in which the user has no part)?

5.4   Experimental Philosophy

Before going into more detail regarding the potential implications of ST, we think it is important to be explicit about our aims here. The main phenomenon of interest here is the way we can describe potential confusions that individual users may experience regarding their agency in relation to actions that are mediated by ST. Hence, we will not (primarily) focus on societal regulation proposals for handling such confusions, for example in relation to establishing accountability or liability. Also, our aim is not to argue that ST must always or often lead to such confusions about agency. We are not making an empirical claim about the magnitude of the risk that symbiotic systems bring. Rather, we are engaging in an analysis of what agency confusions that are due to the use of symbiotic technology could reveal about the standard everyday practice of claiming authorship of actions. Simply put: neurotechnology can create conditions for agency confusions that are realistic, though perhaps in actual practice highly exceptional and philosophically interesting. We are interested in examining the effects of neurotechnology on the experience of ownership of action (e.g., as expressed in verbal reports). Thus, we are using neurotechnology, in particular passive BCIs, as an empirically grounded illustration of (potential) agency confusions. However, we do so not primarily to explore potential ethical risks associated with it, but rather to examine how ST could help to shed light on the processes involved in the SoA. This approach is therefore vulnerable in that it borders between philosophy's traditional (and infamous) thought experiments, where basically anything goes, and what has been called "experimental philosophy" [39], often consisting of data collection (e.g., surveys) in order to investigate how concepts are or could be used under particular circumstances. The list of philosophical thought experiments includes Searle's 1980 Chinese room [40], Jackson's 1982 and 1986 Mary in her black-and-white room [41, 42], or Putnam's 1973 Twin-earth [43] (see also [44]). They are too well-known to require further elucidation.

The logic of the particular type of experimental philosophy we are pursuing here may require some further explanation. With all due respect to the differences, a comparison between neurotechnology and the CERN Large Hadron collider might be informative. Through the collider, exceptional circumstances are created under which certain aspects of matter, otherwise unnoticeable, can be measured and related to theories. In a similar vein, neurotechnology can create special circumstances (although obviously under more stringent ethical conditions) that may help to clarify certain aspects of conceptual usage that would normally remain invisible. Put differently, neurotechnology can enable us to create and/or imagine cases where traditional concepts reveal aspects of their usage "under stress," so to speak, which could lead to a better understanding of when, how, and why such concepts apply, or to reveal cracks in our understanding that may also have consequences for their application under normal circumstances. Hence, neurotechnology may provide opportunities for what we would like to call "empirically guided thought experiments"; extrapolations or imaginations of (near-)possible conditions that, if well chosen, could illuminate our thinking about mind, ethics, law, etc. Still, such uses have to be clearly distinguished, at least in principle, from analyses aimed at establishing factual risks of a technology, which could function as a reason for, e.g., devising codes to regulate practical applications. Other scholars have raised doubts about the application of the concept of action to passive BCIs at all [45]. Our approach, by contrast, deems it possible that actions derived from brain signals via passive BCIs could be called actions and asks what consequences might follow the application of this concept. ST can enable a philosophical investigation of "agency" under exceptional, artificially induced, but empirically informed circumstances.

5.5 Me and My Subconscious Brain States in Action

Discussing the consequences of ST for a user's sense of agency is complicated in part because of terminological issues. In practice, it is often tempting, and sometimes practically almost unavoidable to speak of, e.g., "conscious or subconscious brain states." But this is an imprecise expression [46]. First of all, it is not the brain states that are conscious or subconscious. It is a person, not a brain, that can be conscious or not. Secondly, when persons are conscious of something, it is not of their brain states, as these are introspectively invisible, but of mental states or episodes that supposedly correlate with or are implemented in brain states. Third, although we will focus on brain states, the subconscious information that can be used in ST need not be restricted to neuronal information. One could also consider how the usage of physiological measures (e.g., heart rate, galvanic skin response—GSR) as a source of information could help regard intended actions and examine the implications of a user's sense of agency and responsibility. Finally, it is to be expected that the very diverse types of brain processes underlying the very diverse types of psychological processes (ranging from relatively stable character traits, to changeable behavioral dispositions, to fleeting occurrent thoughts, and from beliefs, thoughts, and intentions to emotions and feelings) do not permit a one-size-fits-all type of analysis. We will not do much more than dip in here and there.

As indicated, we will use ST for a kind of empirically informed or guided thought experiment. STs are philosophically interesting because they provide a technical opportunity to probe the relations between persons and their subconscious brain states. As a starting point, we would like to use the work of Zander et al. [47], involving a task where ST users have to maneuver a cursor towards a specific location in 4 × 4 or 6 × 6 grids. Their application uses event related potentials (ERPs) to collect information about a user's target location (e.g., the upper left corner) in order to direct the cursor movements. The cursor first moves randomly, and when it does not move in the direction of the target, the ERP will display features, e.g., analogous to an Error-Related Negativity (ERN) signal, that over time can be used to infer what the user's target location is. As brain measurements build up and more data points start to support the model of the preferred direction, the cursor will start moving in the right direction. As the authors say:

> In this paper, we demonstrate that by collating passive BCI output and context information, it is possible to develop, step by step, a user model that accurately (…) reveals task-relevant subjective intent. Specifically, we demonstrate that a user model can be developed and used to guide a computer cursor toward the intended target, without participants being aware of having communicated any such information. Using a passive BCI system, the participant's situational interpretations of cursor movements were classified and interpreted, in the given context, as directional preferences [47].

Based on real-time analysis of brain activity, the system established "a user model from which the participants' intentions could be derived" [47].

Zander et al.'s usage of "intentions" is helpful in grasping the contribution of the system to the overall task performance, so we do not want to criticize it (and we have no aspirations in functioning as the word-police anyway), but use it as a starting point for further exploration. Philosophically, it can be useful to consider whether the concept of "intention" genuinely applies here [48–52]. ST provides an opportunity to ask questions about the relation between me and my subconscious brain states in relation to the sense of agency. Do subconscious brain states, as evidenced by the ST's cursor movements, genuinely reflect "my" "intentions"? To what extent is ST deriving intentions, and to what extent could these intentions be considered mine? We will use Dennett's multiple drafts theory as a framework to approach these questions. The point here is not to suggest that our sketchy analyses are correct or even on the right track, but rather to illustrate the kind of questions that arise out of an experimental philosophical usage of developing technology. In addition, note that we expand the case presented by Zander et al. to include cases where the "intentions" are not fixed but can change dynamically, e.g., during the game the users of ST may change their minds about which end location they prefer. That is, we explicitly increase the amount of "thought" in our experimental thought experiment based on Zander's work.

In "Consciousness explained" [53], Dennett suggests his "multiple drafts" model of cognition. In essence, this model holds that many different information processing streams take place in parallel, which are related to aspects of perception, cognition, action, involving multisensory information collection and integration, feedback mechanisms, consequence predictions, memories etc. Only a few of these drafts leave their tracks in actual actions, experiences, memories, or reports of a person. Most drafts disappear without a trace, that is, without a person having had even an inkling that some of his brain processes were dedicated to processing a particular type of information X and Y for such and such a potential action Z. A large part of the brain's information processing that could have surfaced standardly gets irrevocably lost. However, ST enables tapping into parts of the brain machinery involved in devising these multiple drafts. But it is not clear which parts, let alone whether these parts would ever have left traces, *had the technology not been there*. ST is a bit like eavesdropping on a government's planning bureau that is busy developing several elements of various plans in parallel in order to address all kinds of eventualities. But the eavesdropping may take place when it has not yet been determined which elements cohere with certain other parts in an integrated situation assessment that could form the context for developing a coherent action plan. Indeed, it may not even be clear that such a plan will ever be required given the circumstances, let alone that it would actually be acted out. Hence, *it may be the eavesdropping that creates the trace*. Fragments of (potentially different) information processing streams start leaving traces, because the ST records and reflects them. It might be a bit of an exaggeration, but instead of deriving intentions ST might actually be creating them.

An old-fashioned distinction may be of service here. In the early days of the philosophy of cognitive science, Stephen Stich [54] distinguished "lower-level subdoxastic" states from higher-level doxastic ones. Among the latter he counted intentions, occurrent thoughts, beliefs and desires, states of which we can know that we have or entertain them. In contrast, subdoxastic states are isolated from us in the sense that we cannot explicitly use the information they carry in our reasoning or speech, as we have no conscious access to them. Stich concluded that "[s]ubdoxastic states occur in a variety of separate, special purpose cognitive subsystems." Our doxastic states, on the other hand, "form a consciously accessible, inferentially integrated cognitive subsystem." Similarly, Jerry Fodor suggested that

At the very top are states which may well correspond to propositional attitudes [beliefs, desires] that common sense is prepared to acknowledge [...] But at the bottom and middle levels there are bound to be lots of symbol processing operations that correspond to nothing that people – as opposed to their nervous systems – ever do. These are the operations of what Dennett has called sub-personal computational systems [55].

Using this terminology, the effect of ST can be described as "doxizing" subdoxastic states: subdoxastic states are revealed to the subject via ST. The possibility exists that the technology does so prematurely or at least in a way that is different from non-ST mediated action. This means that it is reasonable to ask the question of whether these intentions, if they are such, can genuinely be derived (as opposed to, e.g., constructed) or even whether they are genuinely owned by the person. From this perspective, STs could even be seen as going beyond doxizing subdoxastic states.

Notice that in relation to neurofeedback similar questions can be raised, although a crucial difference is that the user is made aware of their brain activity, whereas in ST it gets translated into an action directly and the user only learns about his inferred intentions from that action *after the act*. Of course, it is true that often we find out about ourselves (intentions, feelings, preferences, etc.) after we act, i.e., by experiencing what we are doing and then reflecting on this (often simply grasping why, or confabulating reasons to explain or justify ourselves). The difference here is that ST introduces a non-personal technological component in that process, which raises the question of whether such acts can be said to be genuinely intended by the person. After all, one witnesses the outcome of a machine's inference regarding one's preferences based on measurements of one's brain states.

Wolkenstein and colleagues [56] have raised a question that is based on this potential of STs to "create" intentions, which the ST user finds out about after he acts. They argue that in cases where a passive BCI acts upon brain signals—and intentions or preferences that are taken to be related to these brain states—it is an open question whether these states represent the true self or a biased, non-unified, version of the self. To illustrate, take Dennett's multiple draft-model again. We saw that a ST could "create" or "construct" intentions out of "non-unified" or "non-ordered" brain processes, which are typically unified and enter consciousness in order to direct a person's actions. When we act we quite regularly assume that what makes us act belongs to our identity, the motivations we have are what makes us the person we are, or so we say. To the extent that STs tap into processes and make some of them action-relevant before they have been processed and made conscious, we would need to know how these "constructed" intentions relate to our identity. Are they the material from which eventually our identity-coherent preferences and intentions are built? Or are they such that it is not clear which of these pre-conscious mental drafts are compatible with our "identity" so that much of the process of making them conscious involves, for example, checking this kind of coherence? Moreover, an answer to these questions entails that we could then judge whether STs reveal a deeper sense of the user's identity (they know more about the user than the user herself does), or if STs potentially distort the user's identity. Note that finding an answer to these questions requires a good deal of research in the philosophy of the mind, something an experimental philosophy approach to STs can certainly elicit.

We do not pretend to have an answer to these questions. Our goal is simply to point out what dimensions of the concept of agency STs touch upon, and what implications in terms of practical conclusions this might have. In summary, ST enables a "tapping into subdoxastic states," presenting information about subconscious brain states in ways that may affect a users' narrative about themselves. Particular types of information that were previously not reported or even reportable can now start to leave traces. Information that has not met the author's explicit stamp of approval, as, e.g., expressed by incorporating it in a report, is now openly soliciting, inviting, or even partially imposing that approval. In addition to investigating the notions of action and intention from a novel perspective, practically important implications for technology-mediated action, e.g., in relation to responsibility and liability can be involved.

5.6   Conclusion: A Call for ST-Based Experimental Philosophy

As we hope to have made clear, ST is philosophically interesting in that it enables subconscious brain states to influence actions in a new, technology-mediated way. Through neuroscience and computer technology, information about subconscious brain states can be made available to support action without the user necessarily being aware of this, or only belatedly becoming aware of it. In one sense such actions can be said to be actions of the users, because their brain states were part of the causal chain leading to the action, and hence at least some part of them was involved in triggering the action. However, questions can be raised as to how this causal contribution can be said to have involved them as genuine agents. We have indicated that it is at least conceivable that the technology may be creating, rather than deriving, involvement by elevating subconscious brain states that could have disappeared without a trace into a cognitive-behavioral episode (doxizing subdoxastic states). We have suggested that this analysis should be seen as an illustration of an experimentally guided thought experiment on the basis of currently researched and suggested symbiotic technology, and not, primarily, as an empirical claim regarding urgent societal implications about existing available applications. Although there may be ethically relevant, societally challenging consequences of ST (with regard to responsibility and accountability), this small plea for experimental philosophy aims to suggest that developing neurotechnology could be used to analyze, reassess, refine, or revise our understanding of, and the conceptual vocabulary for, cognition. We think this opportunity should be used more often.

References

1. Zander TO, Kothe C. Towards passive brain-computer interfaces: applying brain-computer interface technology to human-machine systems in general. J Neural Eng. 2011;8(2):025005. Epub 2011/03/26. PubMed PMID: 21436512. https://doi.org/10.1088/1741-2560/8/2/025005.
2. Krol LR, Haselager P, Zander TO. Cognitive and affective probing: a tutorial and review of active learning for neuroadaptive technology. J Neural Eng. 2020;17(1):012001. https://doi.org/10.1088/1741-2552/ab5bb5.
3. Zander TO, Brönstrup J, Lorenz R, Krol LR. Towards BCI-based implicit control in human–computer interaction. In: Fairclough SH, Gilleade K, editors. Advances in physiological computing. London: Springer; 2014. p. 67–90.

4. Millán JR, Rupp R, Müller-Putz GR, Murray-Smith R, Giugliemma C, Tangermann M, et al. Combining brain-computer interfaces and assistive technologies: state-of-the-art and challenges. Front Neurosci. 2010;4:1–15. https://doi.org/10.3389/fnins.2010.00161.
5. Flemisch FO, Adams CA, Conway SR, Goodrich KH, Palmer MT, Schutte PC. The H-metaphor as a guideline for vehicle automation and interaction. Hampton: NASA; 2003. Technical report NASA/TM-2003-212672.
6. Abbink DA, Carlson T, Mulder M, de Winter J, Aminravan F, Gibo TL, et al. A topology of shared control systems—finding common ground in diversity. IEEE Trans Hum Mach Syst. 2018;48(5):509–25. https://doi.org/10.1109/THMS.2018.2791570.
7. Haller S, Bartsch AJ. Pitfalls in fMRI. Eur Radiol. 2009;19(11):2689–706. https://doi.org/10.1007/s00330-009-1456-9.
8. Poldrack RA. Can cognitive processes be inferred from neuroimaging data? Trends Cogn Sci. 2006;10(2):59–63. https://doi.org/10.1016/j.tics.2005.12.004.
9. Poldrack RA. The role of fMRI in cognitive neuroscience: where do we stand? Curr Opin Neurobiol. 2008;18(2):223–7. https://doi.org/10.1016/j.conb.2008.07.006.
10. Spagnolli A, Conti M, Guerra G, Freeman J, Kirsh D, van Wynsberghe A, editors. Adapting the system to users based on implicit data: ethical risks and possible solutions. Cham: Springer International Publishing; 2017.
11. Gamberini L, Spagnolli A. Towards a definition of symbiotic relations between humans and machines. In: Gamberini L, Spagnolli A, Jacucci G, Blankertz B, Freeman J, editors. Symbiotic interaction symbiotic 2016, Lecture notes in computer science. Symbiotic interaction, vol. 9961. Cham: Springer International Publishing; 2017. p. 1–4.
12. Steinert S, Friedrich O. Wired emotions: ethical issues of affective brain–computer interfaces. Sci Eng Ethics. 2020;26(1):351–67. https://doi.org/10.1007/s11948-019-00087-2.
13. Clausen J, Fetz E, Donoghue J, Ushiba J, Spörhase U, Chandler J, et al. Help, hope, and hype: ethical dimensions of neuroprosthetics. Science. 2017;356(6345):1338–9. https://doi.org/10.1126/science.aam7731.
14. Zander TO, Shetty K, Lorenz R, Leff DR, Krol LR, Darzi AW, et al. Automated task load detection with electroencephalography: towards passive brain–computer interfacing in robotic surgery. J Med Robot Res. 2017;2(1):1750003. https://doi.org/10.1142/s2424905x17500039.
15. Schultze-Kraft M, Birman D, Rusconi M, Allefeld C, Görgen K, Dähne S, et al. The point of no return in vetoing self-initiated movements. Proc Natl Acad Sci. 2016;113(4):1080–5. https://doi.org/10.1073/pnas.1513569112.
16. Zander TO, Gaertner M, Kothe C, Vilimek R. Combining eye gaze input with a brain–computer interface for touchless human–computer interaction. Int J Hum Comput Interact. 2010;27(1):38–51. https://doi.org/10.1080/10447318.2011.535752.
17. Blankertz B, Lemm S, Treder M, Haufe S, Müller K-R. Single-trial analysis and classification of ERP components—a tutorial. NeuroImage. 2011;56(2):814–25. https://doi.org/10.1016/j.neuroimage.2010.06.048.
18. Gerjets P, Walter C, Rosenstiel W, Bogdan M, Zander TO. Cognitive state monitoring and the design of adaptive instruction in digital environments: lessons learned from cognitive workload assessment using a passive brain-computer interface approach. Front Neurosci. 2014;8:385. https://doi.org/10.3389/fnins.2014.00385.
19. Chanel G, Kierkels JJM, Soleymani M, Pun T. Short-term emotion assessment in a recall paradigm. Int J Hum Comput Stud. 2009;67(8):607–27. https://doi.org/10.1016/j.ijhcs.2009.03.005.
20. Kirsh D, Maglio P. On distinguishing epistemic from pragmatic action. Cogn Sci. 1994;18(4):513–49. https://doi.org/10.1207/s15516709cog1804_1.
21. Gallagher S. Philosophical conceptions of the self: implications for cognitive science. Trends Cogn Sci. 2000;4(1):14–21.
22. Saito N, Takahata K, Murai T, Takahashi H. Discrepancy between explicit judgement of agency and implicit feeling of agency: implications for sense of agency and its disorders. Conscious Cogn. 2015;37:1–7. https://doi.org/10.1016/j.concog.2015.07.011.
23. Flemisch FO, Bengler K, Bubb H, Winner H, Bruder R. Towards cooperative guidance and control of highly automated vehicles: H-Mode and conduct-by-wire. Ergonomics. 2014;57(3):343–60. https://doi.org/10.1080/00140139.2013.869355.

24. Synofzik M, Vosgerau G, Voss M. The experience of agency: an interplay between prediction and postdiction. Front Psychol. 2013;4:127. https://doi.org/10.3389/fpsyg.2013.00127.
25. Wegner DM. The illusion of conscious will. Cambridge: MIT Press; 2003.
26. Haggard P, Tsakiris M. The experience of agency: feelings, judgments, and responsibility. Curr Dir Psychol Sci. 2009;18(4):242–6. https://doi.org/10.1111/j.1467-8721.2009.01644.x.
27. Blakemore SJ, Oakley DA, Frith CD. Delusions of alien control in the normal brain. Neuropsychologia. 2003;41(8):1058–67. https://doi.org/10.1016/S0028-3932(02)00313-5.
28. Synofzik M, Vosgerau G. Beyond the comparator model. Conscious Cogn. 2012;21(1):1–3. https://doi.org/10.1016/j.concog.2012.01.007.
29. Gallagher S. Multiple aspects in the sense of agency1. New Ideas Psychol. 2012;30(1):15–31. https://doi.org/10.1016/j.newideapsych.2010.03.003.
30. Gallagher S. The natural philosophy of agency. Philos Compass. 2007;2(2):347–57. https://doi.org/10.1111/j.1747-9991.2007.00067.x.
31. Frith C. Explaining delusions of control: the comparator model 20 years on. Conscious Cogn. 2012;21(1):52–4. https://doi.org/10.1016/j.concog.2011.06.010.
32. Frith CD. The positive and negative symptoms of schizophrenia reflect impairments in the perception and initiation of action. Psychol Med. 1987;17(3):631–48 . Epub 2009/07/09. https://doi.org/10.1017/S0033291700025873.
33. Wegner DM, Wheatley T. Apparent mental causation: sources of the experience of will. Am Psychol. 1999;54(7):480–92. https://doi.org/10.1037/0003-066X.54.7.480.
34. Vlek R, van Acken J-P, Beursken E, Roijendijk L, Haselager P. BCI and a user's judgment of agency. In: Grübler G, Hildt E, editors. Brain-computer interfaces in their ethical, social and cultural contexts. Dordrecht: Springer; 2014. p. 193–202.
35. Haselager P. Did I do that? Brain-computer interfacing and the sense of agency. Minds Mach. 2013;23(3):405–18. https://doi.org/10.1007/s11023-012-9298-7.
36. Wegner DM, Sparrow B, Winerman L. Vicarious agency: experiencing control over the movements of others. J Pers Social Psychol. 2004;86(6):838–48. https://doi.org/10.1037/0022-3514.86.6.838.
37. Fourneret P, Jeannerod M. Limited conscious monitoring of motor performance in normal subjects. Neuropsychologia. 1998;36(11):1133–40. https://doi.org/10.1016/S0028-3932(98)00006-2.
38. Frith CD. Action, agency and responsibility. Neuropsychologia. 2014;55:137–42. https://doi.org/10.1016/j.neuropsychologia.2013.09.007.
39. Knobe J, Nicholas S, editors. Experimental philosophy. Oxford: Oxford University Press; 2014.
40. Searle JR. Minds, brains, and programs. Behav Brain Sci. 1980;3(3):417–24 . Epub 2010/02/04. https://doi.org/10.1017/S0140525X00005756.
41. Jackson F. What Mary didn't know. J Philos. 1986;83(5):291–5. https://doi.org/10.2307/2026143.
42. Jackson F. Epiphenomenal qualia. Philos Quart (1950-). 1982;32(127):127–36. https://doi.org/10.2307/2960077.
43. Putnam H. Meaning and reference. J Philos. 1973;70(19):699–711. https://doi.org/10.2307/2025079.
44. Dennett DC. Intuition pumps and other tools for thinking. New York: W.W. Norton and Company; 2013.
45. Steinert S, Bublitz C, Jox R, Friedrich O. Doing things with thoughts: brain-computer interfaces and disembodied agency. Philos Technol. 2019;32(3):457–82. https://doi.org/10.1007/s13347-018-0308-4.
46. Bennett MR, Hacker PMS. The philosophical foundations of neuroscience. Malden: Blackwell Publishing; 2003.
47. Zander TO, Krol LR, Birbaumer NP, Gramann K. Neuroadaptive technology enables implicit cursor control based on medial prefrontal cortex activity. Proc Natl Acad Sci. 2016;113(52):14898–903. https://doi.org/10.1073/pnas.1605155114.
48. Anscombe GEM. Intention. Oxford: Blackwell; 1957.
49. Bratman M. Intentions, plans and practical reason. Cambridge: Harvard University Press; 1987.
50. Bratman M. Two faces of intention. Philos Rev. 1984;93(3):375–405. https://doi.org/10.2307/2184542.

51. Mele A. Springs of action: understanding intentional behavior. New York: Oxford University Press; 1992.
52. Mecacci G, Santoni de Sio F. Meaningful human control as reason-responsiveness: the case of dual-mode vehicles. Ethics Inf Technol. 2020;22:103–15. https://doi.org/10.1007/s10676-019-09519-w.
53. Dennett DC. Consciousness explained. Boston: Little, Brown and Company; 1991.
54. Stich SP. Beliefs and subdoxastic states. Philos Sci. 1978;45(4):499–518.
55. Fodor JA. Psychosemantics: the problem of meaning in the philosophy of mind. Cambridge: MIT Press; 1987.
56. Wolkenstein A, Jox RJ, Friedrich O. Brain–computer interfaces: lessons to be learned from the ethics of algorithms. Camb Q Healthc Ethics. 2018;27(4):635–46 . Epub 2018/09/10. https://doi.org/10.1017/S0963180118000130.

Brain-Computer Interfaces: Current and Future Investigations in the Philosophy and Politics of Neurotechnology

6

Andreas Wolkenstein and Orsolya Friedrich

Contents

Abstract

Important insights have been generated by ethicists, philosophers, sociologists, lawyers, and representatives from other disciplines regarding the ethics of neurotechnology in general and of brain-computer interfaces (BCIs) in particular. However, since (medical) BCIs have yet to leave the laboratory and the context of clinical studies and enter the "real" world, many important normative questions remain unanswered. In this paper we summarize the main lines of ethical inquiry regarding BCIs, both from the general academic discussion and with a view on the results gained in INTERFACES, an interdisciplinary project on the

A. Wolkenstein (✉)
Institute of Ethics, History and Theory of Medicine, Ludwig-Maximilians-Universität (LMU) München, Munich, Germany
e-mail: andreas.wolkenstein@med.uni-muenchen.de

O. Friedrich
Institute of Philosophy, FernUniversität in Hagen, Hagen, Germany
e-mail: orsolya.friedrich@fernuni-hagen.de

© Springer Nature Switzerland AG 2021
O. Friedrich et al. (eds.), *Clinical Neurotechnology meets Artificial Intelligence*,
Advances in Neuroethics, https://doi.org/10.1007/978-3-030-64590-8_6

normative dimensions of BCIs. Furthermore, we offer our perspective on future research and argue that the ethics of technology should explore decision-making processes by which communities and societies regulate emerging technologies, such as BCIs.

6.1   Introduction

The term "Brain-Computer Interface" (BCI) goes back to a paper from 1970 [1]. Ever since, this technology has seen a massive increase in research activity. However, the first 30 years after the term was coined only saw one or two dozen articles being published per year. This changed right around the beginning of the new millennium. As a brief search for the term "Brain-Computer Interfaces" on PubMed shows, the number of published works jumped from 28 in 2002 to 87 in 2003. A more or less steady increase in the number of articles published after 2003 led to 585 articles being published in 2019. In the early days, scientific research on BCIs prevailed. The first ethics papers did not appear until 2005. This is a well-known phenomenon: the ethical debate on emerging technologies typically sets in after a significant amount of scientific and engineering work has resulted in successful applications of a technology. To this day, important insights have been generated by ethicists, philosophers, sociologists, lawyers, and representatives from other disciplines. However, since (medical) BCIs have yet to leave the laboratory and the context of clinical studies and enter the "real" world, many important normative questions remain unanswered. Moreover, the question of what to do with results gained through work in philosophy and the social sciences is still open. We therefore find it necessary to summarize some relevant discussions and insights that have been generated so far, and to shed light on future issues within BCI research that require further attention. After a very brief overview of the technology known as "BCI," we will lay out the main topics from the ethical assessment of BCIs. We will summarize existing work in applied ethics before summarizing the results from the project "INTERFACES" that has studied BCIs in a number of important ways. Finally, in the last section we offer our perspective on future research and argue that the ethics of technology should explore decision-making processes by which communities and societies regulate emerging technologies, such as BCIs.

6.2   BCIs: Technology and Applications

As Dennis McFarland and Jonathan Wolpaw put it, a BCI is "a computer-based system that acquires, analyzes, and translates brain signals into output commands in real-time" [2]. According to this generic definition, there are three elements that constitute a BCI. The first element is related to the generation (user side) and subsequent acquisition (technology side) of brain signals. There are many ways to generate and acquire brain signals, depending on (a) the type of BCI that is used and (b) the type of brain signal that is detected. The basic distinction divides BCIs into active, reactive, and passive BCIs [3, 4]. In active BCIs, the user has to perform a mental task, which is encoded with a pre-defined meaning (e.g., yes/no). Typically, the mental strategy uses motor imagery (moving one's arm or foot) and encodes the respective movement with a meaning [2]. Reactive BCIs, by contrast, require the user to direct his or her attention to a specific stimulus, which is related to a pre-defined meaning. The brain's reaction to the stimulus is measured and translated into the output connected to the stimulus [2]. Finally, passive BCIs monitor brain activity while the user performs any given task and measures when that brain activity has reached a certain threshold (e.g., drowsiness).

The type of signal that a BCI detects ranges from electric activity (e.g., EEG) to the flow of oxygen in the user's brain (e.g., fMRT). Both invasive and non-invasive methods are used. While invasive BCIs require an intervention, e.g., placing electrodes on top of the cortex, non-invasive forms do not require such an intervention. The electrodes are placed on top of the skull [5].

The second element concerns the measurement and analysis of brain signals [6]. Relevant features are extracted and the information that is needed is analyzed. BCIs often use machine-learning algorithms to obtain the relevant information (e.g., whether a certain mental activity has occurred).

Finally, the information is translated into an output that is used by an external device to perform certain actions (e.g., to pick a letter in an attempt to write something, or to direct a robotic arm) [6].

As this brief explication shows, there are numerous ways to realize a BCI [7]. Most of the current research efforts are put into finding new ways to acquire brain signals (e.g., using fNIRS), developing better algorithms to filter out the information needed, and creating new applications for BCIs. To date, BCIs are used in a number of domains. Medical BCIs were developed to restore lost capacities for communication and transport. Especially persons with physical impairments or patients who have suffered a stroke or spinal cord injury can benefit from the use of a BCI in order to enable communication and improve the rehabilitation process [8–11]. Outside of the strictly medical domain, BCIs are increasingly used in the consumer area [12, 13]. Examples include BCIs for gaming and entertainment purposes. In addition, enhancement through the monitoring of one's brain activity can also be seen as a consumer application.

Exploring the ethical, legal, and social implications of BCIs hinges, to a large extent, on the contexts of their use. There are, however, generic issues that transcend the contexts of use and have been the object of thorough work in applied ethics, to which we will now turn.

6.3   Ethical, Social, and Legal Implications of BCIs: State of the Art

6.3.1   Generic Issues

There are a number of surveys that summarize the views on BCIs by ethicists, BCI professionals, and lay people. Sasha Burwell et al. have studied what issues BCI ethicists raise [14]. In a scoping review, they analyzed a sample (*n* = 42) of bioethics articles and found that there are basically 8 areas of concern that ethics experts have voiced: user safety and risk-benefit analyses, humanity and personhood, stigma and normality, autonomy, responsibility, research ethics and informed consent, privacy and security and justice. Other issues that are regularly, but not as frequently, mentioned include novel domains of application (e.g., military), research funding policy and the need for regulation, as well as responsibility issues regarding the use of machine-learning algorithms.

BCI researchers' opinions on a number of issues regarding the technology have been examined in two studies that both take their data from surveys held at a stakeholder conference that takes place in Asilomar, California each year [15, 16]. Femke Nijboer and colleagues [16] asked experts about their views on issues such as the informed consent process with locked-in patients, risk-benefit analyses, team responsibility, consequences of BCI on patients' and families' lives, liability and personal identity, and interaction with the media. Expert opinion differs in the assessment of specific aspects of these issues, but a majority of them still finds it important to establish ethical guidelines for BCI research and use. Pham and colleagues [15] actually tested a proposal for such guidelines, again with participants from the Asilomar BCI conference. They found broad support for principles that emphasize care for subjects as the prime goal for researchers, modesty regarding the expectations tied to BCI research, a participatory approach to research, a broad, tolerant understanding regarding notions of disability and normality, an acknowledgement of the various relations BCI use might affect, justice with regard to access to BCIs and keeping in mind the broader social impact one's research has.

Lay persons' attitudes are the subject of a recent study by Matthew Sample and colleagues [17]. In the study that comprises answers from over 1400 participants, the authors found two factors that summarize people's concerns regarding BCIs. They call the first factor "concern for agent-related issues," which assembles worries such as "becoming cyborgs, redefining humanity, changing the self, doubting authenticity, defining normality, and enabling unfair enhancement." The second factor includes worries such as "enabling new forms of hacking, limited availability, risking surgical complications, seriousness of device failure, and media hype and inaccuracy" and is called "concern for consequence-related issues."

Some of these issues have also been studied in the project INTERFACES, in which a collaboration between ethicists, sociologists, and lawyers has produced a number of articles. Each of these articles deals with a specific issue and tries to shed light on conceptual as well as normative issues. We shall now turn to this work in more detail.

6.3.2   Results from Conceptual Research from the Project INTERFACES

When looking at BCIs, it is particularly exciting to note that in this neurotechnology we are dealing with a novel form of interaction between humans and machines. The changes people make in the world with BCIs are caused by brain activity alone and do not require any peripheral nerve or muscle activity. The latter can be bridged or replaced by a computer and technical devices [18]. The disembodied character of such a change in the world is the starting point for many philosophical-ethical questions that are related to agency, autonomy, and responsibility.

Another relevant starting point for philosophical-ethical investigations is the fact that the data obtained on brain activity must first be processed and interpreted in a complex algorithmic way in order to provide useful computer instructions. This fact also gives rise to a number of normative questions, such as autonomy, responsibility, and discrimination [19].

The disembodied character of BCI-related changes implies questions regarding the subjective perception of the users, as well as conceptual issues. It has not yet been conclusively clarified to what extent users perceive their effects with BCIs as their own actions and whether they feel fully responsible for them. First empirical results show, however, that BCI users experience themselves as self-determined actors and feel responsible for their actions [20]. However, it is necessary to further empirically investigate whether BCIs can lead to greater distortions of the sense of agency compared to conventional actions. *Sense of agency* usually refers to a pre-reflective or a reflective feeling of the subject that she is causing an action [21]. Errors in the sense of agency could occur in that not only the initiation of the action differs in BCI actions, but also the feedback mechanisms are different. This could lead to a situation in which users do not attribute BCI actions to themselves because they do not feel a sense of their own actions. Conversely, certain events may mistakenly be interpreted by BCI users as their own BCI-generated actions. The possibility of such errors has already been shown in several experimental situations without BCI use [22]. For BCIs, there could be an accumulation of such errors.

Besides such subjective misperceptions of the sense of agency, it is also relevant to discuss if BCI use might have consequences for philosophical concepts of agency and for the standard legal account of actions as bodily movements [23]. In a descriptive manner, active and reactive BCIs can be characterized as a hybrid of mental and bodily action, without involvement of the muscular system [23]. In those BCIs, we can describe a mental action, which is followed by a causal chain and by external effects [23]. Such a hybrid character of action has been shown to be a challenge for the law, as action theory in law is based on a bodily movement requirement [23]. An analysis of philosophical concepts of action and BCIs has shown that (active and reactive) BCI events are actions according to the standard (causal) theory of action, because the event is caused by the right kind of mental state (i.e., an intention) [23]. In contrast, those events that are mediated by passive BCIs cannot be called actions, as they are not caused by the right mental states [23].

While taking a closer look at three different forms of control that are relevant for actions (executory, guidance, and veto control), it can be stated that passive BCIs—contrary to active and reactive BCIs—completely lack executory as well as guidance control [23]. For passive BCIs, there is a conceptual similarity to automatism in everyday life, in which movements are not initiated by a conscious executory command and the person has no conscious guidance control during the movement [23]. It remains an issue for future research, if an installed veto control for the user, which she can operate consciously, would change the conceptual evaluation of passive BCI events from no-action to action.

BCI applications, which do not guarantee conscious action initiation for the user, could be very helpful in making automated processes such as driving a wheelchair easier. For this purpose, it is necessary that the BCI system can predict the intended actions of users with high probability by interpreting brain activity and using machine learning (ML). However, BCIs not only involve algorithms in automated processes, but always require algorithms at relevant operating points (e.g., the extraction and classification of relevant features; transmission of the relevant information to an external device) [19].

Using algorithms and ML in BCIs results in many normative concerns. The inscrutability of the decision-making process and the algorithmic opacity decreases user oversight, comprehension, interpretation as well as control and thus, also trust in algorithms [19, 24, 25]. If a user is unable to understand and to interpret an algorithm and its decision-making logic, the person loses her capacity to control the outcome, which has implications for autonomy [19]. The person will suffer a decrease in autonomy without sufficient knowledge of certain processes of decision-making and without control over data, data acquisition and processing [19]. Discrimination is a further problem related to the inscrutable and inconclusive character of algorithms, as those characteristics make it harder to detect biases, resulting in shortcomings in the decision-making process [19]. Discrimination can lead to less opportunities and less autonomy among discriminated groups and persons and thus, to more or new inequalities in society [19].

BCI use can raise the question not only of agency, but also of the extent to which its use increases or limits the realization of user autonomy. In the medical field, a (re-) opening of (new) possibilities of action seems to be connected with an obviously improved enabling of autonomy, for example by enabling patients to (re-) communicate or to operate a prosthesis independently. There might also be further positive effects though. More or new information about their brain activity could also have a positive effect for users on the realization of autonomy, e.g. if BCI users use the information about their affective states to re-evaluate or re-formulate reasons. Theoretically, positive consequences in the realization of autonomy could also occur, if executory control in action is transferred to the machine after action initiation, thus preventing disturbing influences of human action control from unfolding [26].

However, the modified mode of interaction in BCIs can also lead to the fact that realizing human autonomy becomes more difficult. It is precisely this aforementioned stronger machine control in actions as well as the opaque or even manipulating influence of the machine on human decision-making that can lead to the fact that humans are no longer able to sufficiently realize their autonomy. Especially when recurring information about the user's brain activity in certain situations is algorithmically processed by the computer and presented in similar situations in the future, so that the user only has limited decision-making options, resulting in the possibility of immense losses for the realization of autonomy [26]. The user has less choices and his future is automatically and technologically fixed to his past.

It is particularly important to note that the data obtained through BCIs will probably not be the only data that influence users' decisions. In addition, a wide range of other data (from mobile phones, etc.) about the person is simultaneously collected and could be correlated with each other. This enables complex user profiles. These not only affect the decision-making possibilities and influences on users, but also another aspect of autonomy, namely privacy. Therefore, a further risk of creating such personal data and profiles can be seen in the potential interest of these data by a wide range of institutions, followed by the risk of hacking, violating privacy, and misuse [25].

The previous descriptions make it clear that there are also various questions to be clarified with regard to responsibility when developing and applying BCIs. Following the comments on privacy, it is important to ensure that the individual user or group data on brain activity and correlated mental states obtained via BCI do not fall into the wrong hands and are not used for purposes that the user has not approved or to discriminate against people. However, it should be noted that many people are involved when it comes to ensuring data security and the responsible handling of data. The attribution of responsibility is by no means easy in this context, so that complex regulatory issues must be discussed for this technology. In addition to the question of how to ensure the responsible and secure handling of the data obtained, it must also be discussed whether the users themselves are responsible for the results of their interaction with the machine.

A user can be seen as responsible for possible outcomes with BCIs in terms of legal liability within tort and criminal law [27]. A user can be liable for BCI-mediated actions, but also for omissions that would prevent potential harm [27]. The difficulties of talking about actions and thus responsible actions in some BCI-mediated events were presented before. Further, it can be difficult or even impossible for users to foresee and to prevent many harmful events that the machine causes, as ML and related errors are hard to come by as a human.

An extensive analysis shows, however, that there are no principled objections to imposing civil liability for BCI use that results in foreseeable and non-foreseeable damages to third parties [27]. However, there might be a bigger epistemic gap in criminal responsibility, due to the disembodied character of BCI use and the difficulty of identifying the primary cause of a BCI movement [27]. Such an epistemic gap might cause further difficulties for the presumption of innocence [27].

6.4   A Look Ahead: Focusing on Procedures in the Ethics of BCIs

Assessing BCIs through the ethics lens reveals a number of important issues that are often said to be in need of clarification before BCIs can enter the market. However, it is unlikely that early and far-reaching agreements will be gained concerning these questions. As the academic debate itself shows, there is widespread and continuing disagreement regarding almost all of the issues discussed above. Moreover, when persons outside academic ethics, such as BCI professionals and the public are interviewed, ethical pluralism can be expected and was already shown [17]. In addition to prevailing pluralism in moral concepts in society and among ethicists, the problem of normative assessments of BCIs lies in the inconclusiveness of BCI effects for users and for human action, autonomy, and responsibility. The latter difficulty is partly due to the fact that technology is developing steadily and is not yet fully assessable. In addition, further empirical and philosophical work is needed to better understand and assess the novel interaction between neurotechnology, AI, and humans. Only then can more reliable normative assessments be expected. Parallel to these efforts, it may be useful also to focus on procedural elements in the ethical discussion, especially in times of uncertainty regarding results in technology use.

When it comes to procedural aspects, there are quite a few philosophical precursors to consider. Philosophers in the wake of John Rawls have used an approach in which a procedure is devised whose goal is to find those solutions that no one can reasonably deny, no matter which ethical standpoint they represent. Other ethicists develop their work in close cooperation with practitioners, so that the action-guiding nature of their work is self-revealing. Similarly, some base their work on principles that are well entrenched in our practices or common sense, so that the results reached are thought to be directly action-guiding. However, these solutions will certainly face questions from more theoretically oriented ethicists, and in any case, pluralism and disagreement will remain.

Focusing on procedural elements and considering pluralism as well as continuing disagreement regarding the results of certain technology use, can also mean exploring (political) processes by which varying ethical views can and should come into play.

This proposal implies that, apart from the ethical implications of a certain technology, the processes by which real-world agents debate and find consensus on what to do with this technology need to be explored. This includes a strong focus on processes by which people decide on the ethical nature of using (researching, developing) technology, at least as long as there is a high amount of uncertainty concerning certain outcomes of technology use. In addition, particular decision-making processes for specific questions should be embedded within the larger context of social and political institutions that regulate how societies deal with the impact of emerging technologies. Ethicists should thus reason about how to organize the social and political system in which new questions (and old ones, too) come to the fore and are picked up by relevant institutions. The whole organization of societal, institutional living should be addressed within the ethics of technology as a way to reflect on mechanisms by which innovation on the one hand, and protection of basic values on the other hand are balanced and decided upon. In other words, technology ethics should also be involved or integrated into political and institutional ethics. Heath and colleagues [28] have pursued a similar approach in their attempt to understand business ethics as political philosophy. Another point of reference could be the discussion about "technology governance" that sociologists of technology typically lead [29, 30]. Interrelations between these two fields need to be explored and worked out.

It could appear as though this proposal is based on the idea that it is easier to find common ground on how to shape decision-making processes than on finding the right ethical action-guidance. It might be objected that there is a contradiction involved since one cannot claim that there is insurmountable pluralism regarding ethical evaluations of technology, but not regarding the ethics of institutions. Though we agree with this objection, some arguments can be put forward in favor of putting more efforts into exploring procedural aspects at the current stage of BCI development. First, the suggestion is not meant to substitute work on finding the right ethical action-guidance in BCI use, but rather as an addition in times of uncertainty regarding the results of BCI use. Second, as Stuart Hampshire argues, there might simply be a better chance of finding a consensus on procedures rather than on principles [31]. Third, empirical research on procedural justice has revealed surprising results regarding the power of fair procedures to contain evaluative disagreement. Tom Tyler's work on fair procedures in policing and other domains, for instance, shows how procedural fairness leads to beneficial outcomes, such as cooperation [32] (for an overview, see [33]). Creating fair procedures, by which even normatively disagreeing parties on uncertain technology developments can come to an agreement, remains a promising focus even if uncertainty remains with regard to ethical evaluations in exploring novel technologies like BCIs.

Fair procedures are thus a distinct place of ethical concern, as Ceva thinks [34]. In addition, fair procedures are also constitutive to the acceptability of X and they might also help with the problem that work in applied ethics often lacks action-guidance. Focusing on proceduralist accounts can therefore shed new light on the ethics of BCIs. This line of research can very well be supported by empirical studies on procedural fairness. As mentioned above, fair procedures appear to have a strong influence on people's willingness to accept decisions. In many domains, from the workplace to policing to environmental policy, fair procedures enhance the legitimacy and acceptance of decisions made by authorities. Fair procedures can be considered as realizing demands that the political philosophers have posited as normative requirements. The philosophical exploration of empirical features, by which fair procedures appear to be constituted, remains an open task. At the same time, an application of fair procedures to the question of technology regulation should complement this endeavor and lead to a fruitful trans-disciplinary examination of how technology regulation can be improved.

6.5   Conclusion

In this article we have summarized the main philosophical and ethical aspects of BCIs that previous work in academic ethics has revealed. Issues such as agency and autonomy are important aspects where BCI use can potentially have beneficial as well as detrimental effects. The evaluation hinges on many technological details, such as whether the BCI is active, reactive, or passive. The danger of misattribution of agency, as well as the legal grasp of actions are factors that depend, to a large extent, on the nature of the BCI. Whether BCIs enhance or hamper autonomy is partly a technological issue (e.g., whether there is a mechanism to establish action control), and partly a conceptual one. Closely related to agency and autonomy is the question of responsibility. Discussing legal responsibility in terms of either civil or criminal law has a huge impact on how we assess the forms of responsibility that the legal system allows. Moreover, both the notion of responsibility and the general problem of empirically assessing future developments and outcomes of BCI use, as well as ethical disagreement regarding those outcomes lead us to the necessity of finding processes by which pluralist views can be reconciled. As a result, we propose a stronger focus on proceduralist aspects, where a focus on fair processes could help overcome times of uncertainty in technological development and its ethical evaluations, and by which ethical disagreement can be contained to a certain extent. In summary, even though the debate on BCIs has seen a lot of progress over the last years, there are many avenues for future work, both within an ethical approach, narrowly understood, and a broader political-regulatory perspective.

Acknowledgments Work on this paper was funded by the Federal Ministry of Education and Research (BMBF) in Germany (INTERFACES, 01GP1622A) and by the Deutsche Forschungsgemeinschaft (DFG, German Research Foundation)—418201802. We would like to thank Dorothea Wagner von Hoff for proofreading the article, Meliz-Sema Kaygusuz and Bernadette Scherer for formatting.

References

1. Vidal JJ. Toward direct brain-computer communication. Annu Rev Biophys Bioeng. 1973;2(1):157–80.
2. McFarland DJ, Wolpaw JR. EEG-based brain-computer interfaces. Curr Opin Biomed Eng. 2017;4:194–200.
3. Zander TO, Krol LR. Team PhyPA: brain-computer interfacing for everyday human-computer interaction. Periodica polytechnica electrical engineering and computer. Science. 2017;61(2):209–16.
4. Zander TO, Kothe C. Towards passive brain-computer interfaces: applying brain-computer interface technology to human-machine systems in general. J Neural Eng. 2011;8(2):025005.
5. Graimann B, Allison B, Pfurtscheller G. Brain-computer interfaces: a gentle introduction. In: Graimann B, Allison B, Pfurtscheller G, editors. Brain-computer interfaces: revolutionizing human-computer interaction. Berlin: Springer; 2010. p. 1–27.
6. van Gerven M, Farquhar J, Schaefer R, Vlek R, Geuze J, Nijholt A, et al. The brain-computer interface cycle. J Neural Eng. 2009;6(4):1–10.

7. Rao RPN. Brain-computer interfacing: an introduction. Cambridge: Cambridge University Press; 2013.
8. Chaudhary U, Birbaumer N, Ramos-Murguialday A. Brain–computer interfaces for communication and rehabilitation. Nat Rev Neurol. 2016;12(9):513–25.
9. Salisbury DB, Parsons TD, Monden KR, Trost Z, Driver SJ. Brain–computer interface for individuals after spinal cord injury. Rehabil Psychol. 2016;61(4):435–41.
10. Maksimenko VA, van Heukelum S, Makarov VV, Kelderhuis J, Lüttjohann A, Koronovskii AA, et al. Absence seizure control by a brain computer interface. Sci Rep. 2017;7(1):2487.
11. McFarland DJ, Daly J, Boulay C, Parvaz MA. Therapeutic applications of BCI technologies. Brain Comput Interfaces. 2017;4(1–2):37–52.
12. Blankertz B, Acqualanga L, Dähne S, Haufe S, Schultze-Kraft M, Sturm I, et al. The Berlin brain-computer interface: progress beyond communication and control. Front Neurosci. 2016;10:1–24.
13. Blankertz B, Tangermann M, Vidaurre C, Fazil S, Sannelli C, Haufe S, et al. The Berlin brain-computer interface: non-medical uses of BCI technology. Front Neurosci. 2010;4:198.
14. Burwell S, Sample M, Racine E. Ethical aspects of brain computer interfaces: a scoping review. BMC Med Ethics. 2017;18(60):1–11.
15. Pham M, Goering S, Sample M, Huggins JE, Klein E. Asilomar survey: researcher perspectives on ethical principles and guidelines for BCI research. Brain Comput Interfaces. 2018;5(4):97–111.
16. Nijboer F, Clausen J, Allison BZ, Haselager P. The Asilomar survey: stakeholders' opinions on ethical issues related to brain-computer interfacing. Neuroethics. 2013;6(3):541–78.
17. Sample M, Sattler S, Blain-Moraes S, Rodríguez-Arias D, Racine E. Do publics share experts' concerns about brain–computer interfaces? A trinational survey on the ethics of neural technology. Sci Technol Hum Values. 2020;45(6):1242–70. https://doi.org/10.1177/0162243919879220.
18. Graimann B, Allison B, Pfurtscheller G. Brain–computer interfaces: a gentle introduction. In: Graimann B, Pfurtscheller G, Allison B, editors. Brain-computer interfaces. Berlin: Springer; 2009.
19. Wolkenstein A, Jox RJ, Friedrich O. Brain-computer interfaces. Lessons to be learned from the ethics of algorithms. Clin Neuroethics Camb Q Healthc Ethics. 2018;27(4):635–46.
20. Kögel J, Jox RJ, Friedrich O. What is it like to use a BCI? – insights from an interview study with brain-computer interface users. BMC Med Ethics. 2020;21(1):2.
21. Gallagher S. Multiple aspects in the sense of agency. New Ideas Psychol. 2012;30(1):15–31.
22. Wegner DM. The illusion of conscious will. Cambridge: The MIT Press; 2002.
23. Steinert S, Bublitz C, Jox R, Friedrich O. Doing things with thoughts: brain-computer interfaces and disembodied agency. Philos Technol. 2019;32:457–82.
24. Mittelstadt BD, Allo P, Taddeo M, Wachter S, Floridi L. The ethics of algorithms: mapping the debate. Big Data Soc. 2016;3(2):2053951716679679.
25. Wolkenstein A, Friedrich O. Novel challenges of consenting to brain-data collection and black-box algorithms in BCI use. Bioethica forum. 2019;12(1,2):38–43.
26. Friedrich O, Racine E, Steinert S, et al. An analysis of the impact of brain-computer interfaces on autonomy. Neuroethics. 2018. https://doi.org/10.1007/s12152-018-9364-9.
27. Bublitz C, Wolkenstein A, Jox RJ, Friedrich O. Legal liabilities of BCI-users: responsibility gaps at the intersection of mind and machine? Int J Law Psychiatry. 2019;65:101399.
28. Heath J, Moriarty J, Norman W. Business ethics and (or as) political philosophy. Bus Ethics Q. 2010;20(3):427–52.
29. Kuhlmann S, Stegmaier P, Konrad K. The tentative governance of emerging science and technology—a conceptual introduction. Res Policy. 2019;48(5):1091–7.
30. Simonis G. Technology governance. In: Simonis G, editor. Konzepte und Verfahren der Technikfolgenabschätzung. Wiesbaden: Springer Fachmedien; 2013. p. 161–86.
31. Hampshire S. Justice is conflict. Princeton: Princeton University Press; 2000.
32. Tyler TR. Why people cooperate. The role of social motivations. Princeton: Princeton University Press; 2011.

33. Bobocel RD, Gosse L. Procedural justice: a historical review and critical analysis. In: Cropanzano RS, Ambrose ML, editors. The Oxford handbook of justice in the workplace. Oxford: Oxford University Press; 2015. p. 51–87.
34. Ceva E. Beyond legitimacy. Can proceduralism say anything relevant about justice? Crit Rev Int Soc Pol Phil. 2012;15(2):183–200.

Pragmatism for a Digital Society: The (In)significance of Artificial Intelligence and Neural Technology

7

Matthew Sample and Eric Racine

Contents

M. Sample (✉)
Pragmatic Health Ethics Research Unit, Institut de recherches cliniques de Montréal, Montreal, QC, Canada

Department of Neurology and Neurosurgery, McGill University, Montreal, QC, Canada

Center for Ethics and Law in the Life Sciences, University of Hannover, Hannover, Germany
e-mail: matthew.sample@ircm.qc.ca

E. Racine
Pragmatic Health Ethics Research Unit, Institut de recherches cliniques de Montréal, Montreal, QC, Canada

Department of Neurology and Neurosurgery, McGill University, Montreal, QC, Canada

Department of Experimental Medicine (Biomedical Ethics Unit), McGill University, Montreal, QC, Canada

Department of Medicine and Department of Social and Preventive Medicine, Université de Montréal, Montreal, QC, Canada
e-mail: eric.racine@ircm.qc.ca

© Springer Nature Switzerland AG 2021
O. Friedrich et al. (eds.), *Clinical Neurotechnology meets Artificial Intelligence*,
Advances in Neuroethics, https://doi.org/10.1007/978-3-030-64590-8_7

Abstract

Headlines in 2019 are inundated with claims about the "digital society," making sweeping assertions of societal benefits and dangers caused by a range of technologies. This situation would seem an ideal motivation for ethics research, and indeed much research on this topic is published, with more every day. However, ethics researchers may feel a sense of *déjà vu*, as they recall decades of other heavily promoted technological platforms, from genomics and nanotechnology to machine learning. How should ethics researchers respond to the waves of rhetoric and accompanying academic and policy-oriented research? What makes the digital society significant for ethics research? In this paper, we consider two examples of digital technologies (artificial intelligence and neural technologies), showing the pattern of societal and academic resources dedicated to them. This pattern, we argue, reveals the jointly sociological and ethical character of significance attributed to emerging technologies. By attending to insights from pragmatism and science and technology studies, ethics researchers can better understand how these features of significance affect their work and adjust their methods accordingly. In short, we argue that the significance driving ethics research should be grounded in public engagement, critical analysis of technology's "vanguard visions," and in a personal attitude of reflexivity.

7.1   Introduction: Waves of Technology (Ethics)

In 2019, the prospect of a "digital society" seems to dominate the collective imagination, both in policy and research circles, as well as in popular media. How can it not, with recent high-profile scandals and media events centered on data and privacy, as when Mark Zuckerberg (Facebook CEO) was summoned to appear before multiple governing bodies, including the US Congress and the parliaments of the EU and the UK? This year's shift in attention towards "the digital" did not happen spontaneously, however. Setting aside the longer (and deeply consequential) history of the Internet and other information technologies, governments around the world have already spent several years devising new initiatives and investing resources under this banner. In 2011, for example, EU member states each appointed "Digital Champions," representatives who were given the mandate to "help every European become digital and benefit from an inclusive digital society" [1]. Now, we see digital agendas everywhere, with an internationally booming private sector dedicated to information technology and the emergence of global collaborations for digital forms of governance; the European Commission has announced a new funding program dubbed "Digital Europe." Canada, with a similar rationale, recently signed onto "the Digital Seven" (D7), joining Uruguay and six other nations in pursuing new technological possibilities and promoting "digital government."

Throughout, the word "digital" might imply that there is some shared artifact or infrastructure being envisioned in all of these initiatives. The word has been used to refer to social media platforms and the Internet, and alternatively, as a label for any electronic devices that rely on programming. Nevertheless, the rhetoric being used here is more accurately summarized in terms of sweeping claims about the transformative and disruptive impact of new technologies, rather than in terms of some particular object. Some claims stress the benefits to healthcare, economic productivity, and a whole range of social practices. Others highlight the tremendous dangers and risks of these technologies: enabling authoritarian practices, threatening privacy and equality, diluting the quality of social interaction, spreading misinformation, and so much more. Who is to be believed? There is certainly a hefty dose of social marketing promoting the value and benefits of digital technologies. But who is to say if those positive impacts will materialize or not in technological form and then in concrete social realities? More to the point of the present paper, what is an appropriate response to these developments from academic or policy researchers given these unknowns?

Faced with this quandary, many observers of sociotechnical change will experience a feeling of *déjà vu*. We have already observed multiple waves of technological rhetoric and corresponding societal and ethical worry, particularly among some researchers and in some segments of the media. Since just the 1990s, we have witnessed the rise (and occasional fall) of numerous big ideas: human genomics, nanotechnology, synthetic biology, neural engineering, big data, blockchain, personalized medicine, precision medicine, and most recently, artificial intelligence (AI). The momentary prominence of digital technologies has even led to combinations with previously promoted categories, like the use of machine learning in optimizing clinical care and in training brain-computer interfaces. Along with these waves, academics have assembled technology-centered specializations, most notably for the present paper, "neuroethics" and "AI ethics." For each of these waves, we are presented with a provocation to inquire: "will this be a good technology or a good development?"

But answering this question and even just bringing clarity to it has proven difficult. Frequently and still today, ethics research on emerging technologies seems to be triggered by hyperbolic technological discourse, with limited critical scrutiny of both positive and negative speculative or promissory claims about emerging technologies [2–9]. Academic articles discussing artificial intelligence as an existential threat, for example, are likely to coincide with front-page articles about Google DeepMind, national leaders' speeches about digital innovation, and protests outside of Microsoft offices in Redmond, WA. Far from questioning the technological promises and worries of the day, ethicists may be among the first to reference and reinforce them through conferences, media appearances, and publications [10 12] as we have seen previously regarding other technologies [13]. A corollary is that these questions and worries have remained academic, with limited genuine public engagement and concern for the impact of technology on everyday life [14–18]. As we will argue, such moments of cultural alignment—between ethics research and society—may create a false sense of significance regarding the objects of ethics inquiry. And captive to this skewed sense of significance and relevance, ethics research may become unfit to answer emerging ethical questions rigorously, let alone foundational questions in ethics. As Rayner (2004) has phrased the question, "why does institutional learning about new technologies seem so difficult?" [19].

In this paper, we aim to make progress on this front with respect to scholarship on emerging technology. We advance a discussion about the ethical *significance* of the digital society, in a way that neither dismisses nor naively embraces its societal prominence in 2019. We understand *significance* as both the ethical importance currently granted to certain technologies and, more critically, the open-ended question of what significance these technologies should have within a broader view on human well-being and flourishing. To this end, we will present some recent developments in AI and neural technology as two case studies, creating a high-level picture of the digital society. We will then analyze this picture using insights from science and technology studies (STS) and from pragmatism. We suggest, ultimately, that the current significance of emerging or speculative digital technologies is underpinned by both sociological and ethical factors; that is to say, the digital society is both a product of self-interested technology promoters and something that can impact the well-being of individuals and society, positively or negatively. This has, we conclude, implications for the practice of ethics research on technology and dictates an interdisciplinary approach that focuses on impacts for individual life and democracy. It means, furthermore, that the significance of any given technology is not a foregone conclusion and that ethics researchers themselves have a role to play in foregrounding some problems over others and in attributing significance carefully.

7.2 The Digital Society Is Here, Again: Parallel Trends in Academia and in Society

From social science and law to philosophy and public policy, a wide range of academic disciplines contribute to scholarship that is labeled as "ethics" or is taken to have an "ethical" dimension of inquiry. Part of the intuitive appeal of doing or applying this type of work, which we will refer to here as *ethics research* (or simply *ethics*) is that it is more meaningful than the popular discourses on passing trends and national fads associated with technoscience. Unlike the undisciplined gaze of the public, the lone academic mind or research team can filter out meaning from mere noise. Or so we might think. Yet, there is reason to think that our scholarship on technology maps quite closely onto broader societal trends. Focusing here on two cases (i.e., neural technology and AI), we can see general parallels between academic and societal attention given to emerging or speculative technologies, with only indirect links to actual harms and benefits as experienced by actual (as opposed to hypothetical) persons. As we will argue, the mirroring of academia and broader society is not necessarily or entirely problematic [20], but demands attention in order to properly understand and respond meaningfully to the ethical significance of the digital society.

7.2.1   Neural Technologies

Consider first the case of brain-computer interfaces (BCIs) and, more broadly, neural technology. First devised in the 1970s, neural devices now take many forms, either wearable or implanted, experimental or widely available, and they span medical, commercial, security, and recreational uses [21–24]. As we will discuss them here, these devices are defined by a shared affordance; the user's brain activity can be monitored and/or altered with digital hardware to create another means of interacting with the world. User-controlled BCIs are perhaps most well-known for allowing the user to communicate or control a prosthesis, despite complete bodily paralysis (e.g., due to amyotrophic lateral sclerosis) [21]. Other devices, like implanted deep-brain stimulators and wearable trans-cranial direct current stimulators, alter brain activity directly and have been promised to improve the user's mood or, as widely documented and viewed on YouTube [25], reduce the symptoms of Parkinson's disease. Though these new affordances are covered in the media and have taken on symbolic importance [26], the framing is often primarily positive or promotional rather than critical. Coverage of neuroscientific technologies like BCIs often lacks a balanced consideration of their negative and positive impacts [27, 28].

In the private sector, investment in neural technologies is passing 100 million USD per year, according to one estimate [29]. Meanwhile, Elon Musk (CEO of Tesla and SpaceX) has famously claimed that BCIs are the only way humans can remain relevant and productive in an increasingly automated economy; Musk even started his own BCI company, Neuralink, exhibiting his commitment to that envisioned future [30]. As farfetched and speculative as his reaction may seem, attitudes reported among some publics are not contradictory and indicate a shared measure of optimism when presented with the idea of brain-based devices. A Pew Center poll reports that respondents are simultaneously interested and quite worried about neural technologies that could be used for enhancement [31]. To the same effect, a recent study of public attitudes towards BCI ethics in Germany, Spain, and Canada reports that most respondents expressed moderate to high levels of worry about a wide range of potential ethical concerns of BCI use, but were nonetheless enthusiastic about using neural technology in medical applications [32]. Perhaps unsurprisingly, these trends in broader society are mirrored in research in several domains.

There are many indicators that neural technologies have become an object of concern in academia, with steady growth over the last few years thanks to specialized funding streams in many countries. The United States BRAIN initiative (explicitly oriented towards technology-driven discovery) and the EU Human Brain Project each included dedicated (albeit proportionally small) funding for ethics-oriented research on neuroscience and its applications. Ethics researchers from various disciplines have reacted swiftly to these incentives. The existence of neural technologies has become a prominent, if not questionable, justification of a new field of research [33], "neuroethics," increasingly established since its first conference in May 2002. Non-governmental organizations and research centers, too, have been founded to collect and recognize ethical work related to neural technologies, like the International Neuroethics Society, the Neuroethics Network, Neuroethics

Canada, and the Oxford Center for Neuroethics. The authors of this chapter have also contributed to this phenomenon.

Worries about BCIs reported in this literature range from lack of safety and cost to threats to the user's self-understanding and responsibility; in most cases, these concerns are asserted on the basis of conceptual analysis or philosophical reflection [21]. Despite this tendency, the ethics of neural technology is not "all talk." Underlying the transient and sometimes hyperbolic discourses in academic journals, conferences, press releases, news outlets, and social media, there seem to be some genuinely serious harms and benefits. It is thus imperative to remember that the experiences and well-being of some people have already been impacted, either in the course of BCI research or in its applications. Some participants in BCI research studies, for instance, may gain a new capability to express themselves through participation, using BCI devices to work around communication difficulties. But such research studies sometimes end with either no gain in communicative ability or, in the case of success, with a complete lack of technical and clinical support for continued use [34].

Other notable harms may be related to the interaction between the application of new technologies and political or cultural recognition. Some members of the Deaf community (i.e., individuals who collectively embrace a cultural identity linked to being deaf) have reported that cochlear implants, designed to augment hearing, actually reinforce systematic stigmatization of their bodies and ways of life [35]. Some argue, for example, that hasty promotion of cochlear implants may ultimately preclude the acceptance of Deaf community members as simply different, undermining political obligations to make public infrastructure accessible to them. The general causes of such exclusionary effects are difficult to study, but Deaf activists have supported their critique by citing places and instances in which communities have adapted to more fully support Deaf individuals, rather than requiring the individual to change themselves. A commonly used illustration is the historical example of Martha's Vineyard [36]; in part because of higher local prevalence of hereditary deafness, hearing and deaf individuals alike developed and used sign language, ostensibly a story of greater inclusion of deaf people in public life. This tension between recognition and exclusion has also been reported in reference to BCIs more generally. As part of a multi-stakeholder international deliberation in 2018, potential BCI users and patient advocates reported the ongoing disenfranchisement of individuals due to either the use of stigmatizing language in the promotion of technologies—language that devalues certain types of bodies (e.g., "fixing," "curing")—or the failure to make enabling technologies widely available [37]. In sum, these examples show that the stakes of effective and beneficent neural technology, despite the media hype, are real and deeply consequential in some contexts.

7.2.2   Artificial Intelligence and Machine Learning

Now in 2019, as major brain-oriented government funding initiatives are beginning to sunset, there is space for another platform to represent the digital society. At the moment, this alternative seems to be AI, supplementing front-page media imagery of disembodied brains and wires with screenshots of friendly AI chatbots or schematic representations of neural networks. As neural technologies fade into our collective memory (temporarily or not), we should carefully attend to the way in which AI has newly been constructed as yet another significant technology: sudden and substantial media coverage (despite a longer history), dedicated government funding including ethics research, general awareness among publics, and of course serious stakes and harms for human experience. A schematic comparison with neural technologies reveals the same general pattern at work.

Looking back, AI also has a history that predates its current popularity. While the use of neural technologies often draws on the centuries-old belief that we *are* our brains [38], AI draws on a more recent permutation of the mind-body thesis. The first investments in AI research in the United States were primarily driven by a two-part justification: first, the academic belief that complex human cognition can be modeled and supplemented by computer systems (a form of digital "symbiosis") and, second, the promise to meet Cold War demands for semi-automatic control for precise military decision-making (e.g., with cybernetics) [39]. AI proponents have, however, consistently promoted applications outside of military and defense purposes, applying the computer's ability to solve problems heuristically to medical decision-making [40], logistics [41], translation, marketing, and a range of other uses. Notably, AI has even been applied to BCIs—in the form of adaptive algorithms for the interpretation of neural activity—at least since the early 1990s. These various applications have occasionally coincided with periodic surges in media attention.

In 1984, one commentator in *Science* lauds "exhilarating times" for AI research, and stresses the need to keep promises in line with "the science" [42]. So too, in recent years. At the unveiling of the Stanford Institute for Human-Centered AI, Bill Gates (former CEO of Microsoft) asserted that artificial intelligence is akin to nuclear technology, in terms of its promise and danger to humanity [43]. The *New York Times* has posed the question: "Is Ethical AI even possible?" [44]. These widely reported opinions arrive after a global flurry of publications of national AI strategies, beginning with "A.I. Singapore" [45] in 2017 to the "American AI Initiative" of 2019, which aims to "develop AI in order to increase our Nation's prosperity, enhance our national and economic security, and improve quality of life for the American," but does not directly earmark any additional resources for this task [46]. Reactions collected among some publics track this ambivalent media coverage. In an EU study, 61% of respondents reported positive attitudes towards robots and artificial intelligence, but 88% agree that this technology must be carefully managed to avoid negative impacts [47]. A similar study of Americans reported that approximately 40% of respondents supported the development of AI, with 82% agreeing with the need for careful management [48].

Again, as with neural technology, ethics research on technology has pivoted towards "AI ethics" [49–51]. Part of this is made possible due to dedicated resources from funding agencies and institutions, with some controversial private-public partnerships. Amazon has recently co-funded a US National Science Foundation solicitation for AI and fairness research proposals, matching the government's ten million

USD [52]. Facebook garnered negative press when it announced the establishment of a $7.5 million USD AI ethics institute at the Technical University of Munich, which appeared to some observers as a blatant form of "ethics-washing" for the company [53]. MIT will include AI ethics as a core focus in its newly founded Schwarzman College of Computing. For most of these projects, it is too early to evaluate their outcomes in terms of ethical progress, but there is already a large amount of theoretical content being produced by technology ethics researchers from a variety of disciplines.

Perhaps most remarkable about this global institutional shift to AI ethics is the sheer number of codes of ethics and guidelines that have been produced. In March 2019, a non-exhaustive search yields 9 unique sets of guidelines or principles [54–62], with many more reports and white papers not listed here. The tendency in this genre of document is not to document precise harms or benefits of AI; instead, signatories or authors agree on high-level aspirational goals that should guide the development and use of AI (e.g., beneficence, fairness, democracy, or empowerment) [63]. Some guidelines like the Montreal Declaration and Microsoft's "Approach to AI" are branded explicitly in terms of the private firm or polity in which they originated. Again, as with neural technologies, these transient societal discourses serve many purposes and are only indirectly tied to documented and concrete harms or benefits.

Noteworthy harms exist, nonetheless. Investigative journalists at ProPublica, for instance, report the use of algorithm-based risk assessment software that is more likely to falsely label black defendants as future criminals at almost double the rate of white defendants. This error is consequential, they argue, because judges across the United States can and do cite risk scores during sentencings and at bond hearings [64]. This negative use of AI is sadly not unique and represents a range of practices that motivate the (much more abstract) calls for "fairness" and the avoidance of bias in AI applications [65, 66]. In another high-profile case, an experimental autonomous car designed by Uber struck and killed Elaine Herzberg, a pedestrian in Tempe, AZ [67]. With the absence of regulatory guidance from local or national governments, safety, beneficence, and related ideals for AI, applications demand more than the publication of principles or temporary media attention.

Taken together, where do these considerations leave us with respect to the ethical significance of the digital society? Studied as two exemplars of digital technology, neural technology and AI begin to shed light on some core features associated with ethical significance as it is commonly attributed. In brief these are: periodic surges in media coverage that belie the technology's longer history, dedicated government funding, and some cases of definite harms and benefits related (causally or otherwise) to digital technology. Each feature could be analyzed and evaluated on its own, but together they can be used to think through the nature of significance in ethics research. What lessons do we learn from the fact that this pattern is repeated every few years for each new technological platform? What reaction is appropriate for researchers in ethics of technology?

7.3   Two Wrong Answers: Significance-as-Consensus and Reduction to Hype

Faced with the pattern that we list in the previous section, there are two simplistic interpretations that need to be set aside: *significance-as-consensus* and *reduction to hype*. Though the two perspectives are presented here in their most extreme form and are perhaps not widely held, they facilitate the elaboration of a defendable middle ground (see Sect. 7.4).

First, *significance-as-consensus*: it might be tempting to think that technology promoters, skeptical ethics researchers, and broader society are simply on the right track. That is to say, digital technologies are indeed ethically significant on account of a broad consensus on their importance, positive or negative. After all, why would so many individuals, states, and institutions invest so much energy into talking about them, developing them, and studying them? But this cannot be the whole story. Interpreting significance narrowly as consensus does not tackle the constant shift in framing of technological challenges, from genomics to AI and back again. Moreover, this stance seems to logically imply claims about history of technology that likely no one would accept. For example, *significance-as-consensus* might imply that AI was only societally meaningful in the 1950–1970s and since approximately 2017, during its recent resurgence in popularity. We believe that this is unlikely; the meaningful effects of AI use on communities and individuals likely did not cease in between these periods of attention, though as we will argue this is an empirical question.

At the other extreme, there is *reduction to hype*. More cynical observers will reduce academic-societal attention to technologies as the result of an inevitable boom-and-bust cycle, a predictable outcome of media hype and subsequent disappointment. And the content of hype, as the term's negative valence implies, can be and should be ignored. A commonly cited schematic version of this is the Gartner hype "cycle," which has been criticized for its lack of explanatory power [67]. But as we discuss above, there seem to be real issues present behind the waves of rhetoric and hype, even if the media eventually lose interest. Moreover, the examples we selectively list are just that, examples. Because of the unavoidably limited journalistic and empirical research on the impact of each technology, there remain many unanswered questions about the real utility and value, real-world impact of technology on people's lives, and what matters to people. Reducing technology discourses to mere hype inappropriately discounts this possibility.

7.4   Significant Technologies? Insights from STS and Pragmatism

Ethics researchers can avoid the pitfalls of *significance-as-consensus* and *reduction to hype* to understand how technologies come and should come to be significant. In this section, we will briefly present two tools for critically understanding waves of technological hype and worry: *economies of promising* and the *problems of publics*.

Researchers in the ethics of technology can benefit from learning both, because they are themselves implicated in these processes and these tools can help them reflect critically on their own practices and their understanding of science and technology.

7.4.1   Economies of Promising: Transactions and Vanguard Visions

The first theoretical tool is *economies of promising*. As we present it here, it is a sociological framework and describes how STEM (science, technology, engineering, and mathematics) researchers, funding agencies, and policy-makers use promises—these can be literal promises or related speech acts of prediction or marketing—to create political legitimacy for their activities. In particular, recent literature in STS suggests that the mobilization of resources and the behavior of individual researchers are best understood in terms of high-level interactions between different societal actors. Joly (2010) suggests that the resources given to technoscience are administered according to what promises STEM researchers or tech developers are willing to make, whether to funding agencies or investors [68]. Technoscience is typically practiced in specialized spaces, inaccessible to most citizens, and there is a need to justify the work in terms of what impact it will have outside of the lab. This is a requirement even for researchers who have no intention or capacity to fulfill their promises. Joly labels this system of relationships as an "economy of technoscientific promising" [68].

This fact about the research ecosystem does not imply, however, that those making the promises are wholly dishonest or acting in bad faith. Some sociological research suggests that technoscientific practitioners will often take one of two performative positions with respect to promising [69]. In contexts close to the research (e.g., in a lab meeting), their knowledge of the expansive possibility space and of the myriad technical obstacles results in expressions of deep uncertainty and rejection of guarantees or promises. But when speaking with potential technology users, funders, or taking on the role of the user themselves, scientists and engineers must affirm the imminent utility of their work for concrete practical purposes. Promising in this way does more than just attract resources to one's projects; it yields self-understanding and serves as a source of creativity.

Explained from a slightly different conceptual angle, Hilgartner (2015) suggests that technoscientific researchers must understand and describe their research in terms of an "imaginary" that engages with values present in the surrounding institutional, regional, or national culture [70]. Synthetic biologists in US agriculture, for example, might notice that previously funded projects in other areas of biology are framed in terms of helping fulfill the reputation of the USA as a global leader in food production and food innovation. The synthetic biologists, in response, will frame their own work in this way, borrowing the ends of prior projects, but with a new means. Over time, if this particular project and "vanguard vision" for the technology gains sufficient traction, the discourses around it form a new "sociotechnical imaginary" [71]; this symbolic construct gives meaning to the daily work of technoscientific practitioners who endorse it, but with effects well beyond the lab. This can happen as the imaginary is reinforced and repeated via university press releases, social media, technology blogs, and funding agency reports, among other channels.

Both *economies of promising* and their associated vanguard visions are sociologically important because they help us understand the cultural repertoires from which individual actors craft their behaviors and come to understand themselves. In other words, the action of shared imaginaries resolves the fundamental sociological antinomy of structure and agent [71]. But more to the point of this paper, they also have meaningful connections to ethics-oriented research on technology. First, they highlight that the ethical significance of a given technology is not always an intrinsic property of the objects or infrastructures themselves. The lesson of *economies of promising* is that technology can be made significant by the coordinated efforts of STEM researchers, developers, users, and whoever else has the means and motivation to formulate and promote a "vanguard vision." As a result, we should expect imaginaries to be as numerous as their creators, with all the overlap, contradictions, and conceptual confusion that such multiplicity implies. It means, too, that we can empirically document imaginaries at various stages of uptake, from the first sentence of an unread tech ethics paper (e.g., "Digital technologies are poised to transform society and the lives of disabled people, but may be ethically worrying.") to the internationally broadcasted speech of a prime minister or CEO.

The economy of promising has made its way directly into the work of ethics researchers when they build on existing imaginaries and speculate about the prospective benefits and harms of new technologies. It has been shown how, for example, ethics scholarship has uncritically replicated claims about the transformative impact of various technologies such as deep-brain stimulation [72, 73], 3D bioprinting [74], and cognitive enhancers [75–78]. Given the task of ethics, namely to carefully analyze courses of actions and propose, via deliberation, paths, and scenarios which are promissory of human flourishing (and identify paths which represent obstacles to flourishing), it is not surprising that ethics researchers are concerned about the future implications of technology. By its very nature, ethics is future-oriented although it can build on the past. However, it is in the manner by which claims about the future make their way into ethical analyses about harms and benefits, which summons important theoretical, methodological, and practical concerns.

Second, these imaginaries necessarily feature valuations regarding what is the good life, the good community, etc. although these can be implicit in discourse and practices. These are all evaluable from the standpoint of ethics. Moreover, because imaginaries are the product of intensive work by particular actors, they may not represent the interests of the majority or of the most immediate stakeholders. In the face of historical inequality in society and in technoscientific practices, we cannot expect marginalized individuals or groups to have adequate resources to promote their own sociotechnical visions or to counter unacceptable "vanguard visions," creating a knock-on effect on the process by which technologies become significant. The US rhetoric of global AI leadership, for instance, may have reinforced the significance of AI in the United States without any meaningful input from individuals who have been subject to racial discrimination in the immigration or criminal justice systems (as seen above in ProPublica reporting on sentencing). In this way, the sociological dynamics of significance can short-circuit democratic hopes of representation and could negatively shape our choice of foci in academic ethics research. More problematic is that the power and influence of vanguard visions may imply that other voices be considered as enemies of the nation, of progress, of technoscientific development, and so on. In response, fostering deliberative approaches and dialogue is a clear strategy to offset the narrowness by which vanguard visions and imaginaries are conceived and deployed. This is the object of our next point.

7.4.2   Emerging Publics and Their Problems

The second theoretical tool, *publics*, emphasizes the very real problematic experiences that can sometimes ground discussions of technology. Dewey identifies a public as something that coalesces around shared problems [79]. In his account, the formation of a public starts when a few individuals notice that their chosen forms of life are being affected by some common factor, technoscientific or social, natural or of solely human origin. These individuals can choose to foster a group identity. Once this public has formed, its members can make rights claims or demand political representation in addressing the shared problem. This conception contrasts with the colloquial understanding of "public" as the rather neutral and aggregate sum of individuals in a population. Rather than one singular entity representing the population, there can be multiple publics, just as there can be multiple problems, and with occasional overlap given the existence of multiple interests.

As we discuss above, "the digital society" and its many platforms create a circular or self-referential discourse in many ways, but they also present some very real harms. We know about these harms in part because of the publics that have formed around them. Notable publics of neural technology, for example, include disability activists rejecting the use of cochlear implants to augment hearing. As discussed above, this negative potential of neural technology can be understood in terms of personhood, specifically the experience of being disenfranchised and treated as less than a full political person. In this case, everyday experiences of disability stigma—as echoed by the promotion of the cochlear implant—have been sufficient to cause a variety of citizens to mobilize around this shared threat. Similarly in artificial intelligence, the experiences of unequal treatment in criminal justice, once documented and shared, have galvanized grassroots movements to remedy ongoing government negligence in the proper regulation of algorithm-assisted sentencing. Though these publics do not always form—a significant problem in its own right—their existence can be a focus of ethics research.

Crucially, the assessment of these *publics and their problems* is sociological in nature; in the case of emerging technologies, which publics are formed is predominantly an empirical question, requiring us to go out and look for ourselves and to understand the nature of problematic situations (real or foreseen) based on the experience of those who use/will use technology or be affected by it. We will likely find that some publics (ethicists experiencing a shared need for funding) may have less pressing problems than others (members of the Deaf community who experience stigmatization due to cochlear implants). However, for Dewey, reflection on the formation of publics is an essentially philosophical and political undertaking. Unlike *economies of promising,* the concept of *publics* does not imply transactional, economic, or strategic functions.

Instead, Dewey's *publics* are sites of group intelligence and open-ended possibilities for human flourishing. He understands group deliberation to be a core component of democracy [79, 80], a process of fostering dialogue to enrich our views about present-day problematic situations and, eventually, form a more comprehensive outlook. For pragmatists, this is the sign of a growth in ethical perspectives [81, 82]. The concept of *public,* then, is a reminder that ethics research can and should be an instrument that empowers individuals (grouped as publics) to give their interests a form and to be represented within open-ended (democratic) processes of political legitimacy formation and, in the ideal case, solve shared problems. Conversely, ethics researchers, unaware of the economy of promises and of its impact on them, their practices, their scholarship, their field, etc. (or sometimes deliberately complicit), can perpetuate discourse and practices that alienate everyday experience from the public gaze. The relationship of the ethics researcher with the public is often obfuscated by the servile role of ethics in the economy of promises or the use of ethics knowledge authoritatively to circumvent dialogue [7, 83] or bypass the understanding of perspectives of those concerned [33, 84].

7.5   Significance and Our Responsibility as Researchers

The two tools discussed above—*economies of promising* and *publics and their problems*—reveal the dualistic character of ethical significance in this domain. Specifically, significance in the context of emerging technologies is jointly an ethical category, grounded in the experiences of real people, and a sociological dynamic, which we can embrace or resist. For this reason, we argue that the significance of "the digital society"—it is only the latest in a series of technological waves—demands a nuanced treatment from researchers in technology ethics. It is not enough to simply accept it a priori or to reject it out of hand as mere hype. Building on examples from emerging technologies and on theory, we stress the utility of an interdisciplinary approach to assessing significance. It involves, minimally, three overlapping strategies for ethics research: empirical and deliberative engagement, analysis of dominant imaginaries, and consistent reflexivity in our work.

First, as inspired by Dewey's account of publics, we prescribe engagement with the publics and the problems they face because of technology. Ethics researchers who position themselves as being in service of helping resolve problem situations need to remain true to the nature of the situation, notably that they are not themselves facing the situation and will not provide the answer themselves. That work is for those concerned by the situation. Accordingly, scholars can use empirical or social scientific methods to productively inquire into *which* publics are actually forming, beyond the self-serving discourses perpetuated by technology developers (in promotional roles) or by technology-oriented funding for ethics research. Furthermore, scholars can foster the productive and creative capacities of publics by convening deliberative exercises and events [85–87]. Such events can build on a rich literature on public engagement in sociotechnical change and, as "technologies of humility," can provide a much-needed counterbalance to "technologies of hubris" [16, 88, 89].

The role of the ethics researcher with respect to publics is thus to accompany those concerned in making sense of their situation and helping them find the effective means to transform the situation via open inquiries and ethical deliberation. This does not prevent us from playing an important role, but it should be one in service of the situation so to speak rather than using different (e.g., academic) problem situations to build power and influence in specialist discourses or simply replicate existing valuations without questioning them.

Second, as inspired by work on the *economy of promises,* we recommend critical analysis of dominant sociotechnical imaginaries and vanguard visions motivating our work and implicit in our subject matter. A thorough analysis of imaginaries will require a keen philosophical eye for normative content, documenting and calling out ableist, classist, racist, and other disenfranchising assumptions that animate imaginaries. It will also involve proactively inquiring after the sources of our imaginaries: where did we get them and whose interests does their distribution serve? Which voices and visions are absent? In sum, we propose that the vanguard visions used in ethics research should be recast as "grounded speculation," [4, 90] according to which imaginaries are always situated either in broadly accessible forms of evidence or in deliberative exercises with diverse affected publics.

It has been proposed that greater transparency about different normative and factual (e.g., sociological, scientific) assumptions about speculative claims supporting ethical analyses be recognized as such, including the value attributed to these assumptions within ethical analyses [4]. Furthermore, ethics researchers can adopt methodological measures to cross-check speculative claims by, for example, validating assumptions with literature from different disciplines and adopting a broad perspective to support more comprehensive reflection (see Fig. 7.1) [4]. In this regard, comparing disciplinary frameworks, considering historical knowledge (e.g., evolution of a technology and attitudes towards it), and reflecting on the development of normative approaches towards a given technology are different methodological strategies to instill more objectivity, rigor, and comprehensiveness of ethical analyses which engage with new and evolving technologies.

Our two prescriptions for public engagement and for critical analysis of imaginaries can be recast as a unified intellectual virtue: reflexivity. The dualistic character of significance in this domain entails that responsible scholars are reflexive about their role in perpetuating some problems, some imaginaries, over others. We must be honest about the fact that we rarely conceive of a technological worry *de novo,*

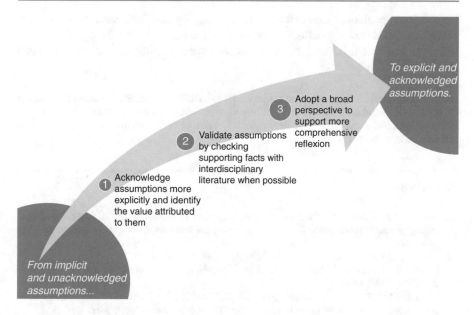

Fig. 7.1 Increasing objectivity and reflexivity in speculation

independent of our cultural moment or of front-page media. The authors acknowledge, of course, that this approach is not wholly new but combines the best features of recent sociological and ethical work on technology, including some of our own recent projects.

7.6 Conclusion

What is the significance of "the digital society" and related technologies such as BCIs and AI? How do we make sense of conflicting utopian, dismissive, and dystopian sentiments about its future? In this paper, we have considered strategic and sociological features of its significance, on the one hand, as well as features associated with actual impact on real-world human experience, on the other. We have shown, by briefly considering two exemplars of digital technology, BCIs and AI, that these two aspects of significance are incompletely but meaningfully connected in society through overlapping technoscientific, social, and academic practices. We show, also, that this entanglement is likely not unique to any one technological platform, but is a common formula across many waves of technological rhetoric.

This social phenomenon can be productively understood through pragmatist and STS resources. However, the understanding that we gain should also be translated into improving our own work on the ethics of technology. For this reason, responsible research in the face of endlessly emerging technologies requires a tailored methodology. Building on previous work, we emphasize the importance of "grounded speculation" and scholarly reflexivity. By taking these lessons to heart, researchers in technology ethics can address concrete problems facing actual communities and move beyond a vague and recurrent feeling of "haven't we been through this before?"

Acknowledgments This project was supported by ERANET-Neuron, the Canadian Institutes of Health Research and the Fonds de recherche du Québec—Santé (European Research Projects on Ethical, Legal, and Social Aspects of Neurosciences), and a career award from the Fonds de recherche du Québec—Santé (ER).

References

1. Digital champions joint mission statement. European Commission. 2014. https://ec.europa.eu/digital-single-market/news/digital-champions-joint-mission-statement. Accessed 28 Mar 2019.
2. Turner L. Bioethic$ Inc. Nat Biotech. 2004;22(8):947–8.
3. Parens E, Johnston J, Moses J. Ethics. Do we need "synthetic bioethics"? Science. 2008;321(5895):1449.
4. Racine E, Martin Rubio T, Chandler J, Forlini C, Lucke J. The value and pitfalls of speculation about science and technology in bioethics: the case of cognitive enhancement. Med Heal Care Philos. 2014;17(3):325–7.
5. Parens E, Johnston J. Against hyphenated ethics. Bioethics forum. 2006. http://www.bioethicsforum.org/genethics-neuroethics-nanoethics.asp. Accessed 28 Mar 2019.
6. Parens E, Johnston J. Does it make sense to speak of neuroethics? Three problems with keying ethics to hot new science and technology. EMBO Rep. 2007;8(1S):S61–4.
7. Evans JH. Playing god? Human genetic engineering and the rationalization of public bioethical debate. Chicago: Chicago University Press; 2002.
8. De Vries R. Who will guard the guardians of neuroscience? Firing the neuroethical imagination. EMBO Rep. 2007;8(S1):S65–9.
9. De Vries R. Framing neuroethics: a sociological assessment of the neuroethical imagination. Am J Bioeth. 2005;5(2):25–7.
10. Forlini C, Partridge B, Lucke J, Racine E. Popular media and bioethics: sharing responsibility for portrayals of cognitive enhancement with prescription medications. In: Clausen J, Levy N, editors. Handbook on neuroethics. Dordrecht: Springer; 2015. p. 1473–86.
11. Caulfield T. The commercialisation of medical and scientific reporting. PLoS Med. 2004;1(3):e38.
12. Caulfield T. Biotechnology and the popular press: hype and the selling of science. Trends Biotechnol. 2004;22(7):337–9.
13. Hedgecoe A. Bioethics and the reinforcement of socio-technical expectations. Soc Stud Sci. 2010;40(2):163–86.
14. Forlini C, Racine E. Does the cognitive enhancement debate call for a renewal of the deliberative role of bioethics? In: Hildt E, Franke A, editor. Cognitive enhancement: an interdisciplinary perspective. New York: Springer; 2013. p. 173–86.
15. Racine E, Gareau I, Doucet H, Laudy D, Jobin G, Schraedley-Desmond P. Hyped biomedical science or uncritical reporting? Press coverage of genomics (1992-2001) in Québec. Soc Sci Med. 2006;62(5):1278–90.
16. Doucet H. Imagining a neuroethics which would go further than genethics. Am J Bioeth. 2005;5(2):29–31.
17. Caulfield T, Condit C. Science and the sources of hype. Public Health Genomics. 2012;15(3–4):209–17.

18. Bubela T, Nisbet MC, Borchelt R, et al. Science communication reconsidered. Nat Biotechnol. 2009;27(6):514–8.
19. Rayner S. The novelty trap: why does institutional learning about new technologies seem so difficult? Ind High Educ. 2004;18(6):349–55.
20. Simonson P. Bioethics and the rituals of media. Hastings Cent Rep. 2002;32(1):32–9.
21. Burwell S, Sample M, Racine E. Ethical aspects of brain computer interfaces: a scoping review. BMC Med Ethics. 2017;18(1):60.
22. Dubljevic V, Saigle V, Racine E. The rising tide of tDCS in the media and academic literature. Neuron. 2014;82(4):731–6.
23. Wexler A. The social context of "do-it-yourself" brain stimulation: neurohackers, bio-hackers, and lifehackers. Front Hum Neurosci. 2017;11:224. https://doi.org/10.3389/fnhum.2017.00224.
24. Clausen J. Man, machine and in between. Nature. 2009;457(7233):1080.
25. Gardner J, Warren N, Addison C, Samuel G. Persuasive bodies: testimonies of deep brain stimulation and Parkinson's on YouTube. Soc Sci Med. 2019;222:44–51.
26. Jasanoff S. Perfecting the human: posthuman imaginaries and technologies of reason. In: Hurlbut JB, Tirosh-Samuelson H, editors. Perfecting human futures. Wiesbaden: Springer; 2016. p. 73–95.
27. Cabrera LY, Bittlinger M, Lou H, Müller S, Illes J. The re-emergence of psychiatric neurosurgery: insights from a cross-national study of newspaper and magazine coverage. Acta Neurochir. 2018;160(3):625–35.
28. Racine E. Neuroscience and the media: ethical challenges and opportunities. In: Illes J, Sahakian B, editors. Oxford handbook of neuroethics. Oxford: Oxford University Press; 2011. p. 783–802.
29. Yuste R, Goering S, Bi G, et al. Four ethical priorities for neurotechnologies and AI. Nat News. 2017;551(7679):159–63.
30. Allen M, VandeHei J. Elon musk: humans must merge with machines. Axios. https://www.axios.com/elon-musk-artificial-intelligence-neuralink-9d351dbb-987b-4b63-9fdc-617182922c33.html. Accessed 28 Mar 2019.
31. Funk C, Kennedy B, Sciupac E. US public wary of biomedical technologies to "enhance" human abilities. Pew Research Center. 2016. https://www.pewresearch.org/science/2016/07/26/u-s-public-wary-of-biomedical-technologies-to-enhance-human-abilities/. Accessed 28 Mar 2019.
32. Sample M, Sattler S, Racine E, Blain-Moraes S, Rodriguez-Arias S. Do publics share experts' concerns about neural technology? A trinational survey on the ethics of brain-computer inter-faces. Sci Tech Hum Val. 2019;45(6):1242–70.
33. Racine E, Sample M. Two problematic foundations of neuroethics and pragmatist reconstruc-tions. Camb Q Healthc Ethics. 2018;27(4):566–77.
34. Klein E, Peters B, Higger M. Ethical considerations in ending exploratory brain–computer interface research studies in locked-in syndrome. Camb Q Healthc Ethics. 2018;27(4):660–74.
35. Sparrow R. Implants and ethnocide: learning from the cochlear implant controversy. Disabil Soc. 2010;25(4):455–66.
36. Oullette A. Hearing the deaf: cochlear implants, the deaf community, and bioethical analysis. Val U L Rev. 2010;45:1247–70.
37. Sample M, Aunos M, Blain-Moraes S, et al. Brain-computer interfaces and personhood: inter-disciplinary deliberations on neural technology. J Neur Eng. 2019;16:063001.
38. Vidal F. Brainhood, anthropological figure of modernity. Hist Human Sci. 2009;22(1):5–36.
39. Edwards PN. The closed world: computers and the politics of discourse in Cold War America. Cambridge: MIT Press; 1997.
40. Topol EJ. High-performance medicine: the convergence of human and artificial intelligence. Nat Med. 2019;25(1):44–56.
41. AI-Powered Supply Chains Supercluster. Government of Canada. 2018. https://www.ic.gc.ca/eic/site/093.nsf/eng/00009.html. Accessed 28 Mar 2019.

42. Waldrop MM. Artificial intelligence (I): into the world; AI has become a hot property in financial circles: but do the promises have anything to do with reality? Science. 1984;223:802–6.
43. Paxton S, Yin W. Bill Gates, Gov. Gavin Newsom speak at unveiling of new human-centered artificial intelligence institute. Stanford Daily. 2019. https://www.stanforddaily.com/2019/03/19/bill-gates-gov-gavin-newsom-speak-at-unveiling-of-new-human-centered-artificial-intelligence-institute/. Accessed 28 Mar 2019.
44. Metz C. Is ethical AI even possible? The New York Times. 2019. https://www.nytimes.com/2019/03/01/business/ethics-artificial-intelligence.html. Accessed 28 Mar 2019.
45. Choudhury SR. Singapore to invest over $100 million in A.I. in next five years in smart nation, innovation hub push. CNBC. 2017. https://www.cnbc.com/2017/05/03/singapores-national-research-foundation-to-invest-150-million-dollars-in-ai.html. Accessed 28 Mar 2019.
46. Accelerating America's leadership in artificial intelligence. White House Office of Science and Technology Policy. 2019. https://www.whitehouse.gov/articles/accelerating-americas-leadership-in-artificial-intelligence/. Accessed 28 Mar 2019.
47. Special Eurobarometer 460: attitudes towards the impact of digitisation and automation on daily life. European Commission. 2017. https://data.europa.eu/euodp/data/dataset/S2160_87_1_460_ENG. Accessed 28 Mar 2019.
48. Zhang B, Dafoe A. Artificial intelligence: American attitudes and trends. Future of Humanity Institute. 2019. https://governanceai.github.io/US-Public-Opinion-Report-Jan-2019/us_public_opinion_report_jan_2019.pdf. Accessed 28 Mar 2019.
49. Anderson M, Anderson SL, editors. Machine ethics. Cambridge: Cambridge University Press; 2011.
50. The ethics and governance of artificial intelligence initiative. 2017. https://aiethicsinitiative.org. Accessed 28 Mar 2019.
51. Bostrom N, Yudkowsky E. Ethics of artificial intelligence. In: Frankish K, Ramsey W, editors. The Cambridge handbook of artificial intelligence. New York: Cambridge University Press; 2014. p. 316–34.
52. Natarajan P. Amazon and NSF collaborate to accelerate fairness in AI research. Alexa Blogs. 2019. https://developer.amazon.com/blogs/alexa/post/1786ea03-2e55-4a93-9029-5df88c200ac1/amazon-and-nsf-collaborate-to-accelerate-fairness-in-ai-research. Accessed 28 Mar 2019.
53. Facebook-funded AI ethics institute faces independence questions. Times Higher Education. https://www.timeshighereducation.com/news/facebook-funded-ai-ethics-institute-faces-independence-questions. Accessed 28 Mar 2019.
54. Floridi L, Cowls J, Beltrametti M, et al. AI4People—an ethical framework for a good AI society: opportunities, risks, principles, and recommendations. Minds Mach. 2018;28(4):689–707.
55. Cutler A, Pribić M, Humphrey L. Everyday ethics for AI design. IBM. 2018. https://www.ibm.com/watson/assets/duo/pdf/everydayethics.pdf. Accessed 28 Mar 2019.
56. Asilomar AI principles. Future of Life Institute. 2017. https://futureoflife.org/ai-principles/. Accessed 28 Mar 2019.
57. IEEE ethically aligned design. IEEE Standards Association. 2019. https://standards.ieee.org/industry-connections/ec/autonomous-systems.html. Accessed 28 Mar 2019.
58. Diakopoulos N, Friedler SA, Arenas M, et al. Principles for accountable algorithms and a social impact statement for algorithms. 2016. http://www.fatml.org/resources/principles-for-accountable-algorithms. Accessed 28 Mar 2019.
59. Montréal declaration for responsible development of artificial intelligence. 2019. https://www.declarationmontreal-iaresponsable.com/la-declaration. Accessed 28 Mar 2019.
60. The European Commission's high-level expert group on artificial intelligence: ethics guidelines for trustworthy AI. European Commission. 2018. https://ec.europa.eu/digital-singlemarket/en/news/draft-ethics-guidelines-trustworthy-ai. Accessed 28 Mar 2019.
61. Artificial intelligence at Google: our principles. Google AI. 2019. https://ai.google/principles/. Accessed 28 Mar 2019.
62. Microsoft AI principles. Microsoft Corporation. 2019. https://www.microsoft.com/en-us/ai/our-approach-to-ai. Accessed 28 Mar 2019.

63. Whittlestone J, Nyrup R, Alexandrova A, Dihal K, Cave S. Ethical and societal implications of algorithms, data, and artificial intelligence: a roadmap for research. Nuffield Foundation. 2019. http://www.nuffieldfoundation.org/sites/default/files/files/Ethical-and-Societal-Implications-of-Data-and-AI-report-Nuffield-Foundat.pdf. Accessed 28 Mar 2019.
64. Angwin J, Larson J, Mattu S, Kirchner L. Machine bias. ProPublica. 2016. https://www.propublica.org/article/machine-bias-risk-assessments-in-criminal-sentencing. Accessed 28 Mar 2019.
65. Calo R. Artificial intelligence policy: a primer and roadmap. UCDL Rev. 2017;51:399.
66. Courtland R. Bias detectives: the researchers striving to make algorithms fair. Nature. 2018;558:357–60.
67. Somerville H. Uber shuts Arizona self-driving program two months after fatal crash. Reuters. 2018. https://www.reuters.com/article/us-autos-selfdriving-uber/uber-shuts-arizona-self-driving-program-two-months-after-fatal-crash-idUSKCN1IO2SD. Accessed 28 Mar 2019.
68. Joly PB. On the economics of techno-scientific promises. In: Akrich M, Barthe Y, Muniesa F, Mustar P, editors. Débordements. Mélanges offerts à Michel Callon. Paris: Presses des Mines; 2010. p. 203–22.
69. Brown N, Michael M. A sociology of expectations: retrospecting prospects and prospecting retrospects. Technol Anal Strateg Manag. 2003;15(1):3–18.
70. Hilgartner S. Capturing the imaginary: vanguards, visions and the synthetic biology revolution. In: Miller C, Hagendijk R, Hilgartner S, editors. Science and democracy: making knowledge and making power in the biosciences and beyond. London: Routledge; 2015. p. 51–73.
71. Jasanoff S, Kim SH, editors. Dreamscapes of modernity: sociotechnical imaginaries and the fabrication of power. Chicago: University of Chicago Press; 2015.
72. Gilbert F, Viaña JNM, Ineichen C. Deflating the "DBS causes personality changes" bubble. Neuroethics. 2018. https://doi.org/10.1007/s12152-018-9373-8.
73. Gilbert F, Ovadia D. Deep brain stimulation in the media: over-optimistic portrayals call for a new strategy involving journalists and scientists in ethical debates. Front Integr Neurosci. 2011;10(5):16.
74. Gilbert F, Viaña JNM, O'Connell CD, Dodds S. Enthusiastic portrayal of 3D bioprinting in the media: ethical side effects. Bioethics. 2018;32(2):94–102.
75. Wade L, Forlini C, Racine E. Generating genius: how an Alzheimer's drug became considered a 'cognitive enhancer' for healthy individuals. BMC Med Ethics. 2014;15:37.
76. Forlini C, Racine E. Added stakeholders, added value(s) to the cognitive enhancement debate: are academic discourse and professional policies sidestepping values of stakeholders? AJOB Prim Res. 2012;3(1):33–47.
77. Forlini C, Racine E. Disagreements with implications: diverging discourses on the ethics of non-medical use of methylphenidate for performance enhancement. BMC Med Ethics. 2009;10(1):9.
78. Racine E, Forlini C. Cognitive enhancement, lifestyle choice or misuse of prescription drugs? Ethics blind spots in current debates. Neuroethics. 2010;3(1):1–4.
79. Dewey J. The public and its problems. Denver: Swallow Press; 1927.
80. Pappas GF. John Dewey's ethics: democracy as experience. Bloomington: Indiana University Press; 2008.
81. Pekarsky D. Dewey's conception of growth reconsidered. Educ Theory. 1990;40(9):283–94.
82. Gouinlock J. Dewey's theory of moral deliberation. Ethics. 1978;88(1977–1978):218–28.
83. Evans JH. A sociological account of the growth of principlism. Hastings Cent Rep. 2000;30(5):31–9.
84. Fiester AM. Weaponizing principles: clinical ethics consultations & the plight of the morally vulnerable. Bioethics. 2015;29(5):309–15.
85. Racine E. Éthique de la discussion et génomique des populations. Éthique publique. 2002;4(1):77–90.
86. Doucet H. Les méthodes empiriques, une nouveauté en bioéthique? Revista Colombiana de Bioética. 2008;3(2):9–19.

87. Doucet H. Le développement des morales, des législations et des codes, garder le dialogue ouvert et la conscience inquiète. In: Hébert A, Doré S, de Lafontaine I, editors. Élargir les horizons: Perspectives scientifiques sur l'intégration sociale. Sainte Foy: Éditions Multimondes; 1994. p. 135–41.
88. Jasanoff S. Technologies of humility: citizen participation in governing science. Minerva. 2003;41(3):223–44.
89. Doucet H. Le développement de la génétique: quelle tâche pour l'éthique? Isuma. 2001;2(3):38–45.
90. Voarino N, Dubljević V, Racine E. tDCS for memory enhancement: analysis of the speculative aspects of ethical issues. Front Hum Neurosci. 2017;10:678. https://doi.org/10.3389/fnhum.2016.00678.

Brain-Computer Interface Use as Materialized Crisis Management

8

Johannes Kögel

Contents

Abstract

Making use of the conceptual pair of crisis and routine according to the understanding of Ulrich Oevermann, brain-computer interface (BCI) use can be described as materialized crisis management. When using a BCI, users find themselves in a critical position, which is a situation that requires a decision within a limited period of time. This situation is constantly repeated, turning BCI use into a back-to-back decision-making scenario. BCI use can be described as a permanent crisis that needs to be dealt with. The ultimate goal is to establish a routine of action. Conceptualizing BCI use as a form of crisis management may

J. Kögel (✉)
Institute of Ethics, History and Theory of Medicine, Ludwig-Maximilians-Universität (LMU) München, Munich, Germany
e-mail: johannes.koegel@med.uni-muenchen.de

© Springer Nature Switzerland AG 2021
O. Friedrich et al. (eds.), *Clinical Neurotechnology meets Artificial Intelligence*,
Advances in Neuroethics, https://doi.org/10.1007/978-3-030-64590-8_8

foster an understanding of the BCI user on a procedural level and may suggest improving suitability for daily use and user comfort by optimizing ways of routinization in technology use.

8.1   Background

What exactly does the user of a brain-computer interface (BCI) do? When looking at the test person in the BCI laboratory that we visited, all we can really see is a person sitting in front of a screen and wearing an electroencephalography (EEG) cap with cables attached that lead to some technological device (receiver). He does not speak, and shows no signs of movement. He just stares at the screen. On the screen is a matrix displaying the letters of the alphabet. Some of the letters are flashing in quick succession. After a few moments a letter appears at the bottom of the screen in the output line. Every few moments another one is added to the former. The only instruction that the test person was given was to focus on the letter that he wants to choose.

If you had not been able to see what was happening on the screen, you would think that the test person was not doing anything at all. In fact, you are merely assuming that the output on the screen is somehow related to what the test person is doing. Knowing that the computer will select a letter anyways after a certain period of time has elapsed, it may well be that whatever the test person is doing or thinking is completely unrelated from the output on the screen.

Besides the user, there are two persons from the scientific research team present who are observing various computer screens. Their screens show the test person's brain activity as measured via the EEG. One can recognize event-related potentials that can be associated with decision-making processes of the user.

Except for the output on the screen of the user, the neurofeedback, the readouts of the brain activities, statistical data and computational models, not much is known regarding the processing of BCI use. Most empirical research focuses on usability aspects and asks test persons how they perceived various technical and functional components and how to improve the technology. Research studies on what is actually happening during BCI use, taking into account the only person introspectively involved, the user, are rather sparse. Looking at BCI users' accounts, I shall theorize in this chapter how BCI use may be described and understood. Applying the terms of crisis and routine as outlined by Ulrich Oevermann [1, 2] to interviews with BCI users may render some insights that can be of use for future BCI developments.

8.2   Sociology of Brain-Computer Interfaces

When it comes to brain-computer interfaces, medical and technical researchers as well as philosophers have done a lot of work, but almost no research has been done by sociologists. Scoping the range of studies that have been conducted by means of social research methods, it immediately becomes apparent that mostly quantitative methods are being employed [3]. The few studies that employed qualitative research methods, such as qualitative interviews or focus groups, mainly apply these methods to their respective research projects, but are not social science studies. They are what could be called "descriptive ethics," which aim for "a snapshot of opinion" [4]. They may also be called descriptive medicine or descriptive technology research. Studies that applied a social research design are scarce.

To my knowledge, there is but one comprehensive sociological study on BCIs that has been published. Melike Şahinol [5] has done extensive ethnographic fieldwork on BCIs.[1] Her focus was on the mutual adaptation process between human and machine. According to Şahinol, BCI applications lead to a socio-biotechnical adaptation process based on a cybernetic principle that results in a "techno-cerebral subject," which constitutes some form of a bio-technical cyborg. Şahinol suggests that neuroscientists detach brain processes from body experiences by trying to isolate cerebral entities from bodily movements in order to delegate cerebral motion activities to the computer. This could possibly result in changes in the user's self-experience and reflexivity, and an altered "Leiberfahrung." Accordingly, the worldview inherent in neuroscience is one of a Cartesian "cerebro-centrism" that reduces humans to cerebral functions.[2]

In a recent interview study with BCI users, we identified some central themes of BCI use from the users' perspective: feeling like the agent of BCI actions, opportunities for social participation, and ways and modes of self-definition [9]. As in most cases the BCI output is in accordance with the users' intentions, the users have the feeling that they are the cause and the authors of BCI-mediated actions, whether this may be playing a game, operating a communication program, or steering a robotic arm. Beyond these direct action outputs, users also cherish the indirect effects of BCI training, which include having a meaningful task, contributing to science and technological progress, displaying BCI skills and BCI-generated output to the public, and raising awareness. In addition, through the course of BCI training users are handed opportunities as well as challenges that affect the way they see themselves. It may allow them to get back to activities that they were previously used to pursue, lead to new experiences, result in learning new skills, or cause a (re)thinking about their body and their self in different ways. The BCI may be seen as the medium that allows for these things to happen. In the following, I want to focus on the BCI as an installation or network consisting of all the components mentioned above: the hardware (screens, computers, cables, EEG cap), the software, persons present, and proceedings and events happening (flashing of characters on the screen, appearances of

---

[1] In this respect, it should be mentioned that she observed BCI applications that can be called "hybrid BCIs" [6, 7]. These were BCIs that were supplemented by external brain stimulation devices via deep brain stimulation or electrocorticography or where electromyographic signals were read out in addition to electroencephalographic signals.

[2] The advent of a "cerebral subject" which is characterized by the "creed" that we as humans are essentially our brains has also been stressed by Vidal and Ortega [8] who observed postulates of the neurosciences and their influence way beyond its disciplinary and scientific realm.

letters in the output line, brain activity displayed). As noted above, to the observer it appears as though not a lot is happening. Hence, we must ask the users what exactly is happening when they use a BCI. As a consequence, this research endeavor is concerned with the procedural level of BCI use from the users' perspective. According to their account, I shall argue, BCI use can be described and understood as materialized crisis management.

8.3 Methods

In order to explore the field of BCIs, our research team visited several BCI laboratories; we witnessed some BCI training, tested the technology ourselves, and talked to BCI researchers and developers. In addition, we conducted qualitative interviews with nine BCI users that had or have been using BCIs in experimental studies for which they had been selected due to particular physical impairments. As the research question at hand aims for the procedural level of BCI use from the point of view of users, the following examination will focus on the interview material.

The participants of the interview study varied regarding age, gender, physical condition, and BCI application used. The semi-structured interviews contained questions regarding the personal background of the users and how they came into contact with BCIs, their experiences with the technology, action-related aspects of its use, and a general assessment of BCIs.[3]

Three interviews were conducted face-to-face at the users' homes. One interview was held in written form, due to the participant's speech impediments. Two further participants with oral speech impediments answered questions in written form, combined with oral communication at their homes. Three interviews with participants from other countries were conducted via video calls online. All interviews declared informed consent for their participation in the interview study. The interviews were transcribed verbatim and analyzed using the coding procedure of Grounded Theory methodology [10, 11].

The notion of crisis management was gained by abduction: it emerged as a fit to the data in embracing various themes such as concentration issues, challenges posed, confrontation with unfamiliar puzzles, ways of overcoming handling failures, and experiences of frustration as well as achievement and satisfaction.[4] Turning

---

[3] A detailed portrayal of the participants and the interview guide are available in Kögel et al. [9].

[4] The concept of abduction stems from Charles Sanders Peirce and is utilized in Grounded Theory as an additional logic of reasoning. By means of a "mental leap" [12], data that seem to make no sense or to be contradictory are accounted for. Finding a theoretical explanation, however, requires going back to the data and examining whether they fit the explanation one arrived at [11]. The concept of crisis management appears to account for two findings:

1. Even though the users manage to operate a BCI (and generally feel like being the agent of BCI-generated actions), BCI use poses various hardships and challenges that users are not always able to deal with successfully.
2. While the outside perspective often renders the perception of a flawless proceeding, the introspection of the users often differs: the BCI output may be brought about by a lot of mental

to theories of crises, Oevermann provides a categorization of crises while putting them in contrast to the notion of routine, a concept that easily qualified as a salient theme within the interview material. Turning back to the data, occurrences and descriptions of routinization were highlighted and categories of crises were identified, modified and supplemented (e.g., by the category of mini crises).

8.4   BCI Use as Materialized Crisis Management

Examining the users' accounts, BCI usage can be characterized by permanent decision-making, strenuousness, moments of surprise and frustration, and the endeavor for finding action patterns. To conceptualize BCI use as crisis management we need to clarify some concepts first.

8.4.1   What Is a Crisis?

"Krisis," in its original meaning, refers to a "judgement" or "decision." Oevermann [1, 2] distinguishes three kinds of crises: traumatic crises, crises of decision-making, and crises of leisure.

A traumatic crisis is the result of a sudden, unforeseen event (a "brute fact") that appears like a stroke of fate, e.g., a car accident or winning the lottery. It is impossible not to react to this kind of crisis. Every reaction is already part of overcoming or coping with this kind of crisis.

A crisis of decision-making is due to an uncertain situation regarding how to act, which requires a decision within a limited period of time. Its origin is not external, but stems from the perception or vision of several potential future scenarios. The practice of life itself constitutes these crises due to the contingent nature of actions (by constantly constructing different opportunities and alternatives to act). There is a necessity to make a decision and it is impossible not to make a decision.

A crisis of leisure can manifest itself in the contemplation of an object leading to contradicting perceptions. When engaging with something in a purposeless way, new details of the object and new angles may emerge that can lead to an altered perception, which appears as a crisis. This type of crisis is brought about by the subject itself. Regarding the cause of the crisis and its perception, the crisis of leisure differs from the traumatic crisis, which is induced through an external cause and is experienced as some form of stress.

All three kinds of crises have in common that they are the opposite of the concept of routine. A crisis is opposed to the "pattern of smooth action processing" [1]. A crisis is therefore the disruption of patterns of flawless action processing. Crises can

---

effort and struggle or the generated output is actually not congruent with the user's intentions.

As a consequence, the concept of crisis management seems fit to explain the inner world of BCI usage according to the interviewees' accounts.

be generalized as the questioning of routines of perception, whereby these crises can be either positive or negative.

To manage (from Latin "manus": hand) means as much as "to direct (by hand)", for example a horse. The best literal translation may therefore be "to handle." I therefore want to use the word in its meaning of "to handle or direct with a degree of skill" or "to work upon or try to alter for a purpose" and not of "to succeed in accomplishing" [13].

Crisis management then refers to the necessity of dealing with a certain situation (crisis) in a more or less structured or organized way, in contrast to an approach of ignorance or a passive way of letting things go.

8.4.2   BCI Training as Repetitive Action Crisis

The different kinds of crises pointed out by Oevermann can be applied to the experience of BCI users.

8.4.2.1 Traumatic Crises
BCI users often find themselves in a situation where they are aware of what they are physically able to do and what not. They are aware of the difference that they hope the BCI can evoke. Some are in need of a wheelchair due to a car accident; some see themselves confronted with motion and speech impediments because of a stroke they suffered. As a consequence, they may experience something that we commonly refer to as a life crisis. They experience a situation where everyday routines and routines of movements have become obsolete. Here, a BCI may provide opportunities to facilitate rehabilitation and exercise new routines of movement. Individuals with degenerative diseases regard BCIs as technology that may become relevant for them once they reach a certain level of impediment. For some individuals with advanced amyotrophic lateral sclerosis (ALS), BCIs are already their only way of maintaining communication. It can be proposed that BCI users became users of this neurotechnology due to traumatic crises that they experienced in the past.

8.4.2.2 Crises of Decision-Making
A crisis of decision-making comes to the fore at the very beginning or, to be accurate, prior to BCI use, when participants of BCI studies have to make a decision on whether they want to apply for the experimental BCI studies in the first place. However, they had no say in whether or not they were selected as study participants. This decision is made by the persons in charge of the studies based on the results of various test trials. They also have no say in getting a home-based BCI. A home-based BCI has only been installed in very few singular cases.

Regarding BCI training as such, it may be described as a series of crises of decision-making. Each and every command that is being sent (and transmitted neuronally) from the BCI user during BCI training can be seen as the user's response to a crisis situation. The user finds herself in a situation where she is supposed to perform a certain task. This task can consist of having to select a certain item, such as a character in a P300 BCI scenario, or to imagine moving a particular limb in a BCI run by motor imagery. The user is in a wanting situation and needs to decide somehow. In addition, most BCI applications are set up in a way so that brain activities are measured and transformed into computer commands in certain time intervals or react according to any changes in brain activity. As such, it is also not possible to not make a decision. The computer always translates non-decisions into particular choices and makes a decision anyway. Thus, BCI use is crisis management on a meticulous level: within second or minute intervals (even when no definite time intervals are programmed) you are expected to make choices and decisions, which require conscious thought coordination that accompanies BCI use.

Having to decide on every stimulus given or presented to you in the BCI training, i.e., making back-to-back decisions, is the exact opposite of an action routine. You cannot simply do stuff, you consciously need to bring about every single command. BCI training may therefore be described as a permanent or repetitive action crisis. Wolfgang, a study participant with muscle atrophy, states:[5]

> I had to concentrate on what to choose. That requires – that wasn't easy. You really need to focus. You mustn't get distracted somehow (Wolfgang).

Wolfgang was working with a BCI based on the mental strategy of selective attention, where he needed to choose an item on a monitor within certain time intervals. The difficulty hereby does not lie in the complexity of the task, but in its practical execution:

> It's not very difficult as such. To get the idea of it, to comprehend its principle, is not difficult. But it is hard to execute without any mistakes—that it always recognizes what you want it to do. But as such it's not difficult to comprehend (Wolfgang).

Keeping concentration levels high while making choices—in this case item selections—over long periods of time becomes quite a challenging and tiring task.

This kind of crisis of decision-making differs in some regards to other crises of decision-making, like the one described above about applying for a BCI study (I guess this is the type of decision-making crisis that Oevermann had in mind). The crises within BCI training are under an even more rigid time regime due to the artificial structural framing. The BCI study is an experiment where users are given particular tasks, which they are supposed to perform within a given period of time or within regular time intervals. This setting and the instructions given are something external that determines your decision-making.

Robert, who has been paraplegic since having had an accident, has trained with a BCI-steered exoskeleton. He reports on how changing the way you think when using a BCI as compared to former times, proves to be quite challenging:

> So it was not easy, because when you walk, you don't think yourself. I guess, myself before the accident, I was not thinking about right leg moving, it was automatic (Robert).

---

[5] The following citations stem from the interview study mentioned above [9]. Citations from interviews that were conducted in German have been translated by the author. Names have been pseudonymized.

Moving your legs, e.g., to walk, used to be something that happened automatically. Now Robert needs to consciously think about moving a leg.

The routinization of thought processes and mental strategies, which can take a lot of time and training sessions, is being compromised by new tasks and instructions or a change in setting. Tasks are made more difficult or are applied to new devices or applications. Also, different body postures may lead to a mental re-direction:

> Of course, you have for people like me who have not been standing for years, you have to be able to refocus (Robert).

Robert refers to a change of position from his wheelchair to the exoskeleton. Sitting in his wheelchair, i.e., a position Robert has been used to for the past few years, BCI training worked fairly well. In a standing position, however, like for the training with a BCI-steered exoskeleton, he felt strange and insecure. He needed to readjust his mental strategy and, consequently, his training output deteriorated.

8.4.2.3 Crises of Leisure

Strictly speaking, crises of leisure are not possible in BCI training in that you are always supposed to perform certain tasks. Hence, a free-floating mind is not part of the training. There may be, however, moments in the interviewees' accounts that point in this direction. That is when you look at BCI training more globally, i.e., look at BCI training in total and abstracted from single commands. Robert explains what it felt like to work with a BCI:

> Well, quite strange at the beginning then because you said well, it's a kind of a dream, but it doesn't mean anything. After a while, surprisingly a little bit to have this image and perception of yourself and representation of your, for me, legs, it was something, how can I say, which helped me a little bit to see that my body was not working, but was still there and fully there. It was something very important for me, because I know I could see my legs before, but it was something that didn't exist in myself. It was a very strange feeling before the experience and therefore I am more reconnected maybe not with a computer but with myself. Although it doesn't move, it doesn't move, but it belongs to me [laughs], but it's strange what I say (Robert).

Robert talks about a feeling of strangeness, which is based on the discrepancy between what he can see and what he can feel. This eventually leads to new body- and self-awareness. While occupied with this self-observation—in a rather purposeless way—in particular the observation of his body or his legs to be precise, or of the virtual avatar he is training with (this is meant by "representation"), his perception is changing. What emerges is some sense of being "reconnected" to his body and himself.

8.4.2.4 Traumatic Crises 2.0

At the same time, the acquisition of new perceptions and abilities or regaining former abilities can cause traumatic crises. New perceptions of one's body or one's self that BCI training can bring about can lead to immediate reactions, which can be either positive or negative.

As mentioned above, Robert switched from BCI training in a wheelchair to BCI training in an exoskeleton, which needed him to "refocus." His first time in the exoskeleton was described as follows:

> There was this excitement to be again in this position. Something quite strange also linked to that [...] I had trouble the way you could look in my wheelchair. If you are in front of me, I can see you but sometimes you are not - you don't have this panoramic view, if I can say. At the beginning, since I haven't been standing for years or ages, I was frightened [...] I was stressed, really stressed, scared and they were really there to help me, to say, secure the - they attach things in the exoskeleton in case I was rolling or whatever. Everything was in a perfect thing but it was me to be stressful, but after some weeks, it was fine (Robert).

This literal change of perspectives to the by now unfamiliar "panoramic" view of a standing position caused powerful effects. He felt excitement, panic, and fear— until he got used to it after a couple of weeks of training.

For Nicole, who was diagnosed with spinocerebellar ataxia, BCI training changed her self-image.

> Oh, God, it really changed my self-image. It changed, as I said, the empowerment. The feeling, 'I did this, look what I can do.' It helped me realize that-- I have a saying up on my wall, 'You are more than the body you live in.' I just realized the truth of that statement, that my brain was the most important part of me, and that working meant I could do a lot (Nicole).

The realization that her brain was still in control over her body when connected to a BCI gave rise to a feeling of empowerment and a changed self-image. The empowerment was due to being able to do things she thought she cannot do anymore. Nicole also worked with a robotic arm that she steered via BCI which enabled her to feed herself again, for example. This changed self-image is due to a focus on her brain or mind. Your self is not necessarily about your body. Your self (as the mental part of a person) is the essential part of you, which is embodied in your brain. Nicole shifts the center of her self-understanding from her body to her (mind-embodying) brain.

Stefan, who was born with generalized dystonia, was hoping that a BCI could allow him to operate a personal computer on his own. However, his spasms did not allow continued BCI use and thus he was not accepted for the BCI study. As such, hopes of gaining more autonomy can lead to disappointment.

These traumatic events may not be as sudden and unforeseen as Oevermann described in traumatic crises. Nevertheless, these situations brought about strong emotions as immediate consequences that needed to be dealt with in some way. Robert got used to the upright position and after a while the initial fear made way for excitement. Nicole's feeling of empowerment led to a rearrangement of her self-image, realizing that her brain was the most important part of her. Stefan had to continue his search for appropriate technology and found an eye tracker with which he is able to control a computer by himself.

8.4.2.5 Mini Crises

In BCI training, there are situations that I would like to refer to as "mini crises" as they are limited to disturbances that can occur during BCI training, mostly as a consequence of failed BCI commands. They are basically a follow-up of crises of decision-making and can have an impact on future decisions.

Executing commands can be very exhausting. It takes a lot of energy and concentration or requires some tedious trial-and-error, such as the task of thinking about nothing. Rudi, who is tetraplegic due to a car accident, has been training on a BCI game in which you have to imagine different things in order to send the right command. Usually the commands were given by imagining body movements, but for one command you had to think about nothing, a task that was experienced as particularly challenging.

> To NOT think in particular is ULTRA hard to do and as good as impossible, because everybody is thinking somehow all the time (Rudi).

Robert faced a similar problem. When asked by the principal investigator of the BCI study what he has learned, he responded:

> Oh, I learned to think about nothing. But it's not easy to say 'I am not going to think about anything except my body.' It's something that [laughs] is progressive learning, if I can say. Because at the beginning I was saying 'I should have come with my wife, go with my wife to yoga' (Robert).

Robert did not have the particular task to think about nothing, but he refers to the need of not getting affected by anything. Getting distracted can distort the BCI output. To focus on one's body and body movements and to ignore everything else takes a lot of practice. This includes cutting off thoughts about things one could have or should have done, like in Robert's case going to yoga with his wife. Rudi is also familiar with this challenge:

> I still remember the first training. I had got to know someone and she told me before the training 'It's not working out between us'. So my thoughts were completely somewhere else and then trying to relax – hence, the training session was shit, logically. These influences do exist. Or when my pet died. These influences do exist as they do for everyone. That's also the thing when it comes to not thinking. Your mind wanders automatically. That's how it is (Rudi).

For most users, BCI training is quite a strenuous, tedious trial-and-error process. Tasks such as "not thinking" make it even harder.

Furthermore, unsuccessful BCI actions can lead to frustration. Mrs. Edlinger, who has amyotrophic lateral sclerosis, describes the conditions that are necessary for BCI use:

> Everything needs to be adjusted perfectly and the electrodes must be placed correctly, otherwise success is low or it doesn't work at all. Then you are frustrated (Mrs. Edlinger).

Sometimes the BCI reads out a command that is different from the one you had in mind. This is reported as being frustrating as well, in particular because only the user knows that they actually intended to perform a different task. For external observers this remains opaque. In addition, you are not supposed to give way to your frustration as this may negatively affect the next commands.

> I was NOT allowed to get upset, because that also would have been a signal and that would have influenced the training. You are NOT allowed to have emotions, none whatsoever. You completely have to be dead inside (Rudi).

This means you are not allowed to express your emotions, as this will have an impact on the next commands. Controlling one's emotions is quite a difficult task. This "imperative of no emotions"—in the case of transgression—can also give rise to a mini crisis. Doubts have a similar effect on BCI training. When asked if unsuccessful BCI actions come as a surprise to him, Robert reports:

> That's quite difficult for me, because sometimes I thought I could really focus and concentrate and it was quite a surprise. For instance, you don't answer properly that means [I] was thinking on the right and it was the left and when it was like that I shouldn't have—I should really not care about, because if I care it's even worse. That means that the next round, it goes wrong also. Next try sorry, it goes wrong because I have too much focus on why, why, why so I try to say, 'Okay, I don't know', and then also the two [researcher] and [researcher], they said [...] 'It can happen'. So I try not to figure out why because I don't – frankly I don't understand. If I try to really focus too much, this has an impact on the second try or the third try (Robert).

If you are not able to contain your emotions or distracting thoughts this mini crisis can prolong for the next two or three commands, as described in the cases of Rudi and Robert.

Mini crises can be added to the traumatic crises, crises of decision-making, and crises of leisure. The various crises and the ultimate goal of BCI training, which is to achieve some level of routinized BCI actions, can be seen in Fig. 8.1.

The stages or levels of these crises do not always co-exist; they can all occur and have been reported by BCI users that participated in the interview study, but do not necessarily occur for all of them. The exception is the crises of decision-making. These have been defined as situations that require a timely response in the form of a decision being made and therefore describe the very content and challenge of BCI use.

As the described crisis management becomes visual in the form of the BCI apparatus with all its hardware components, the crises not only take place in the mind, but also have a physical, haptic component. At the same time, a BCI works via "discrete data" [5], readable brain activity that is captured graphically or otherwise, i.e., concrete neurons firing, which is displayed in some way and eventually results in some form of feedback, may it be visual, auditory, or haptic. BCI use therefore becomes materialized crisis management. The BCI is not viewed here as a medium that enables and gives rise to agency, participation, and self-definition. Instead, the focus is on how the BCI is used taking all its components, including its physical parts and their contribution to BCI proceedings, into account.[6]

---

[6] The ambiguous character of technology has been stressed in particular by system theoretical approaches of technology [14, 15]. Jost Halfmann [14] stressed that technology works as a medium

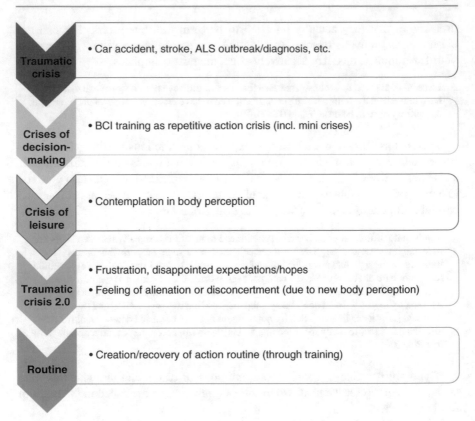

Fig. 8.1 Crises in BCI use

8.4.3   Routinizing BCI Use

The goal of BCI training is to attain a level where BCI actions become routine. In contrast to BCIs with an implemented mental strategy of selective attention where—taking the example of a spelling program—"you really had to fight for every single letter" (Wolfgang), routinization in BCIs that work with motor imagery seems possible. Although Robert was quite exhausted at the beginning of the BCI training study, "after a while, I got used to and it was better, better and better, I should say" (Robert).

The new routine can materialize into regaining an ability, whereby the goal is to get back to a level of automatization. As mentioned above, moving his legs used to be "automatic" for Robert, whereas now he needed to consciously bring about each motion.

---

as long as it works flawlessly. Its side as an installation becomes apparent especially when technology fails or malfunctions. Then it can become a source of irritation to the social world. This can also be tracked in our example of BCI use. Failures of the technology, however, are not only traced in the material components, but also in cognitive mechanisms.

Nicole describes the process of routinization by learning to use the BCI "instinctively":

> Yes, in the beginning, I was concentrating very hard. Then I was thinking, 'Move right, move right, move right'. As I stopped concentrating so hard, I just trusted that I knew how to do it and started to just do it without thinking about how I was doing it, then it became much easier. I realized I didn't have to try as hard as I was trying, that my brain had not forgotten. // Actually, it worked best when you just acted by intuition? // Yes, exactly when I let instinct take over (Nicole).

For Neil, who is tetraplegic since he had a car accident, BCI use felt kind of easy from the very beginning, whereby the research team must have managed the difficulty of the tasks in a well-tuned manner:

> So even on the very first day it worked and it wasn't like something that I really had to try hard for. For the first couple of weeks they, you know, they start you off pretty slow […] we just did the progression slowly enough that it never really felt like I had to really strain myself to do these tasks, I just, kind of, worked at it (Neil).

At the end, Neil entered a stage where BCI use became something that he was simply capable of:

> I always say it was really cool at first and now it's just second nature. Like, it's just, it's a thing I can do at this point. I mean, it's still really cool, but it's just a thing I can do and it's pretty easy to do. It's just almost like I could always have done it before (Neil).

Now he even has the feeling that he can anticipate what is expected of him:

> There's that kind of stuff where, you know, just even instinctually, not even, not even consciously I'll adapt to what the computer ends up expecting. You kind of just develop a rhythm for that kind of stuff (Neil).

This routine portrayed by Neil is similar to what has been described as the transparency of activities of technology use. This notion of "transparency," understood in a Heideggerian sense, has been outlined by Grübler and Hildt [16]. It relates to processes in which you can just focus on the task and be ignorant of the actual handling of the tool. For example, you can think of what you want to draw and what your drawing is supposed to look like when using a pen instead of having to focus on how to grasp the pen and move it correctly to perform the task.

Neil and Nicole explained this by illustrating how they work with a robotic arm via BCI:

> [I]t's like, yes if you're going to reach for, you know, your coffee cup on your table, you don't think 'Oh I need to lift my shoulder up, unbend my arm and reach with my hand and then, and then I close my hand'. You just think, you know, 'I'm going to go grab my cup' and that's just how it works (Neil).

> If I looked at the robotic arm and I wanted to go right, I didn't have to beg, 'Oh, go right, go right, go right.' I would just look at which way I wanted it to go, then my brain automatically made it go that way. Then it became a very natural, very easy thing to do (Nicole).

While Nicole talks about moving the arm in a particular direction, Neil refers to the task of grabbing a cup. In both examples, they talk about simply performing whatever they intended to do. They just had to make up an intention to perform a particular task and the BCI performed that very task. It was not necessary to consciously accompany each and every intermediate step.

It should also be noted that the two BCI users who talked about BCI use as being relatively easy and transparent, Neil and Nicole, both had implants which most probably allow for better readouts in comparison to non-invasive BCIs. For them, a routinization in their BCI use is clearly apparent. They describe their BCI use as "rhythm," "natural", "second nature", and as something they do "instinctually."

Walter, a study participant with muscle atrophy, used the term "automatically self-determined" (Walter) when referring to non-disabled people, while disabled people may be dependent on other people on particular occasions. This also may be seen as a long-term goal of the BCI technology: to achieve a level of action routine in BCI use that allows for "automatic self-determination" in situations where its users used to have to depend on others.

8.4.4   Recommendations for Future BCI Development

Given these circumstances, the most important point in order for BCIs to establish themselves as sustainable technology may lie in mastering routinized BCI handling. There have been attempts to induce *flow* to BCI use [17], which still needs to go a long way.[7] This has been tried for motor imagery settings, but attentive selection settings seem to be impervious to such attempts. While aiming to install an action routine in BCI use, some preferences regarding particular BCI settings seem to be self-evident:

First, motor imagery as a mental strategy seems to be preferable over selective attention. It may not be as robust and easy to learn from the very beginning, but it can turn into a seemingly effortless practice for trained users.

Second, it seems to be more rewarding to be able to act on one's own time and level of difficulty, as has been demonstrated by the flow experiments [17]. A setup that does not work with predetermined time intervals takes some pressure off of the users, and when the BCI only reads out intended brain activity and does not read out all brain activity, frustration for unwanted BCI commands can be avoided.

Third, other sources for frustration could be eliminated if the BCI would be able to discern affective brain activities and process them in a way that does not impede BCI use. For Artificial Intelligence, Marvin Minsky postulated that it needs to learn how to deal with emotions [20]. BCIs probably need to learn to at least detect emotional brain states and to suspend them from other brain activity. Trying to suppress your affective side—because it can have an effect on BCI use—can be tedious and difficult, and can lead to frustration whenever emotions negatively impact BCI

---

[7] Research has shown that "fluency of actions selection" has a positive effect on one's sense of agency [18], while tedious or difficult decision-making has a negative effect [19].

outcome. It appears to be quite difficult for a user to fully embrace this technology while at the same time it cannot accomplish an essential part of what makes us human.

There are many more features that can be programmed or set up in an action routine fostering way, which BCI engineers and technicians are better informed of.[8]

8.5   Conclusion

The BCI laboratory is not only the venue for BCI training, but also denotes—from the experiential side of the user—the space of crisis management in a nutshell.

The conceptualization of BCI use as materialized crisis management is proposed as one possible way to make the very process of using a BCI intelligible. Regarding BCI use as crisis management can foster the understanding of this social process and could increase sensitivity for what BCI users experience or are going through when using a BCI.

Having to make back-to-back decisions under time pressure for long periods of time is not something that we normally encounter in everyday life. Since we are not used to it, we experience tasks like the ones performed via the BCI as tedious and exhausting. To implement a way of automatic or routinized BCI operation may therefore be seen as the crucial point in turning BCIs into a viable and sustainable technological device. In order for technology to be incorporated into our "life-world" as the "unexamined ground of the natural world view" [22] it needs to allow for a routine way of acting.

Acknowledgments This article is based on an interview study which was funded by a grant from the German Federal Ministry of Education and Research (01GP1622A) within the ERA-NET Neuron program. The funders were not involved in data collection or analysis or writing of this article.

References

1. Oevermann U. Sozialisation als Prozess der Krisenbewältigung. In: Geulen D, Veith H, editors. Sozialisationstheorie interdisziplinär: Aktuelle Perspektiven. Stuttgart: Lucius & Lucius; 2004. p. 155–81.
2. Oevermann U. "Krise und Routine" als analytisches Paradigma in den Sozialwissenschaften. In: Becker-Lenz R, Franzmann A, Jansen A, Jung M, editors. Die Methodenschule der Objektiven Hermeneutik. Wiesbaden: Springer VS; 2016.
3. Kögel J, Schmid JR, Jox RJ, Friedrich O. Using brain-computer interfaces: a scoping review of studies employing social research methods. BMC Med Ethics. 2019;20:18.
4. de Vries R. How can we help? From "sociology in" to "sociology of" bioethics. J Law Med Ethics. 2004;32:279 92.
5. Şahinol M. Das techno-zerebrale Subjekt: Zur Symbiose von Mensch und Maschine in den Neurowissenschaften. Bielefeld: Transcript; 2016.

---

[8] For a discussion of ability expectations that users need to bring to BCI training, see Kögel and Wolbring [21].

6. Pfurtscheller G, Allison BZ, Brunner C, Bauernfeind G, Solis-Escalante T, Scherer R, Zander TO, Müller-Putz G, Neuper C, Birbaumer N. The hybrid BCI. Front Neurosci. 2010;4:1–11.
7. Müller-Putz G, Leeb R, Tangermann M, Höhne J, Kübler A, Cincotti F, Mattia D, Rupp R, Müller K-R, del R. Millán J. Towards noninvasive hybrid brain–computer interfaces: framework, practice, clinical application, and beyond. Proc IEEE. 2015;103(6):926–43.
8. Vidal F, Ortega F. Being brains. Making the cerebral subject. New York: Fordham University Press; 2017.
9. Kögel J, Jox RJ, Friedrich O. What is it like to use a BCI? – insights from an interview study with brain-computer interface users. BMC Med Ethics. 2020;21:2.
10. Strauss A, Corbin J. Basics of qualitative research: grounded theory procedures and techniques. Newbury Park: SAGE; 1990.
11. Charmaz K. Constructing grounded theory. London: SAGE; 2006.
12. Reichertz J. Abduction: the logic of discovery of grounded theory. In: Bryant A, Charmaz K, editors. Handbook of grounded theory. London: SAGE; 2007. p. 214–28.
13. The Merriam-Webster.com Dictionary. Manage [Internet]. Merriam-Webster Inc.; 2019. https://www.merriam-webster.com/dictionary/manage. Accessed 19 Dec 2019.
14. Halfmann J. Die gesellschaftliche "Natur" der Technik. Eine Einführung in die soziologische Theorie der Technik. Opladen: Leske und Budrich; 1996.
15. Japp KP. Die Technik der Gesellschaft?. Ein systemtheoretischer Beitrag. In: Rammert W, editor. Technik und Sozialtheorie. Frankfurt a.M: Campus; 1998. p. 225–44.
16. Grübler G, Hildt E. On human–computer interaction in brain–computer interfaces. In: Grübler G, Hildt E, editors. Brain-computer-interfaces in their ethical, social and cultural contexts. Dordrecht: Springer; 2014. p. 183–91.
17. Mladenović J, Frey J, Bonnet-Save M, Mattout J, Lotte F, editors. The impact of flow in an EEG-based brain computer Interface. Proceedings of the 7th graz brain-computer Interface conference 2017; 2017.
18. Chambon V, Sidarus N, Haggard P. From action intentions to action effects: how does the sense of agency come about? Front Hum Neurosci. 2014;8:320.
19. Sidarus N, Haggard P. Difficult action decisions reduce the sense of agency: a study using the Eriksen flanker task. Acta Psychol. 2016;166:1–11.
20. Minsky M. The emotion machine: commonsense thinking, artificial intelligence, and the future of the human mind. New York: Simon & Schuster; 2006.
21. Kögel J, Wolbring G. What it takes to be a pioneer: Ability expectations from brain-computer interface users. Nanoethics. 2020;14:227–39.
22. Schütz A, Luckmann T. The structures of the life-world. Evanston: Northwestern University Press; 1973.

The Power of Thoughts: A Qualitative Interview Study with Healthy Users of Brain-Computer Interfaces

9

Jennifer R. Schmid and Ralf J. Jox

Contents

Abstract

Brain-Computer Interfaces (BCIs) use the power of thoughts. They detect brain activity to control external devices such as neuro-prostheses or personal computers. The goal of this study was to explore the experiences of healthy persons using BCIs in various applications. Based on maximum variation sampling, 24 qualitative interviews were conducted with healthy BCI users, such as neuro-gamers, pilots, users of consumer BCIs, as well as BCI developers and research-

J. R. Schmid
Institute of Ethics, History and Theory of Medicine, Ludwig-Maximilians-Universität (LMU) München, Munich, Germany
e-mail: jennifer.schmid@campus.lmu.de

R. J. Jox (✉)
Institute of Humanities in Medicine and Clinical Ethics Unit, Lausanne University Hospital and University of Lausanne, Lausanne, Switzerland
e-mail: ralf.jox@chuv.ch

© Springer Nature Switzerland AG 2021
O. Friedrich et al. (eds.), *Clinical Neurotechnology meets Artificial Intelligence*,
Advances in Neuroethics, https://doi.org/10.1007/978-3-030-64590-8_9

ers. Our findings indicate that human-machine interaction is influenced by BCIs in a novel and unique way. The success of BCI use was highly linked to motivation and duration of training. Discomfort was mainly associated with the time-consuming procedure of electroencephalography (EEG). Moreover, cognitive exhaustion by BCI use was reported. Most participants expressed being puzzled and fascinated by BCIs, showing a high level of ambivalence regarding BCI technology.

9.1   Introduction

A brain-computer interface (BCI) is a technical device that uses the power of thoughts. It detects and processes brain activity in order to modulate external devices, therefore allowing a person to give commands or communicative expressions without the need to use any other part of the body. Thus, external devices like protheses, wheelchairs, or personal computers can be controlled mentally [1, 2]. Three main categories of BCIs exist: active, reactive, and passive BCIs [1, 3]. Active BCIs permit intentional actions using mental strategies like motor imagery [1]. In this way, specific brain activity patterns are intentionally evoked by the person in order to effectuate specific external actions via a computer [1, 3]. Here, the most frequently used detector is electroencephalography (EEG) [2, 4]. In contrast, in reactive BCIs users direct their attention to external stimuli like flashing lights, which allows them to indirectly control the computer output [1, 5]. Finally, passive BCIs simply monitor mental state changes, such as attention, fatigue, or workload, and modulate external devices based on predetermined thresholds of these brain activities [6, 7]. Here, mobile BCI systems with dry electrodes (such as *Muse* and *Emotiv Epoc*) are often used [7].

In the last decades, BCIs were mainly developed for medical purposes: replacing, restoring, and enhancing functions of the central nervous system [2]. BCIs are a promising tool for neuro-rehabilitation after stroke or as assistive devices for patients suffering from locked-in-syndrome or amyotrophic lateral sclerosis (ALS) [8–11]. However, non-medical BCI applications are becoming more and more popular. A few years ago, two American visionaries - Mark Zuckerberg (Facebook) and Elon Musk (Neuralink) - announced the development of BCIs exclusively for healthy people [3]. Furthermore, non-medical applications of BCIs in planes and automobiles are becoming more and more realistic in the near future [12–14]. In addition, military BCI applications and BCI games are attracting peoples' attention [15–20].

A broad range of questions arise with this new and complex forms of human-machine interaction. How do healthy BCI users describe this special experience? What about ethical and legal considerations, e.g., regarding control and responsibility for BCI-modulated actions? So far, empirical BCI studies have mainly assessed the effects of the technology and the acceptance of it by users [21–24]. Only very few socio-empirical studies have looked at the fears and expectations of users in more depth [21, 25, 26]. To our knowledge, this is the first qualitative interview study that explicitly focuses on the experiences and perspectives of healthy BCI users.

9.2 Methods

9.2.1 Study Design

A qualitative interview study was conducted from July to December 2017 ($n = 24$). All study participants had to have significant experience using BCI technology for non-medical purposes. Before starting, all interviewees consented to being recorded with an audio device. The interviews were either conducted face to face ($n = 18$) or via phone/Skype ($n = 6$). The study received approval by the research ethics committee of the Medical Faculty of the Ludwig-Maximilians University Munich (No. 17-238).

9.2.2 Recruitment and Sample

According to maximum variation sampling, our goal was to recruit a heterogeneous sample of study participants (female = 5; male = 19) [27]. The subjects' age varied between 25 and 56 years (mean = 35 years). Prior BCI experience differed from amateur knowledge (one-time BCI use; $n = 11$) to expert knowledge (professional career in BCI context; $n = 13$). The sample included aviation pilots who have used BCIs in the context of a study ($n = 7$), BCI researchers who have used the technology themselves several times ($n = 10$), BCI soft-/hardware developers ($n = 2$), people who play BCI games ($n = 3$), and university students with experiential knowledge of BCIs from study participation ($n = 2$). Nineteen interviews were conducted in German and 5 interviews in English. The interviews lasted 39 min on average. Table 9.1 provides an overview of our sample.

9.2.3 Data Collection and Analysis

Twenty-four guided interviews were conducted. On the one hand, the interview guide provided clear orientation and structure [28]. On the other hand, it offered a high level of flexibility focusing on participants' individual perspectives [28–30]. The interview guide was semi-structured according to the following topics:

- Concrete BCI use (e.g., number of times, duration, mental strategies, training)
- Ethical aspects of BCI technology (e.g., feeling of control, responsibility)
- Subjective perception of BCI use (e.g., expectations, hopes, fears, future potential)

Table 9.1 Sample - naïve user: amateur knowledge due to one-time BCI use ($n = 11$); expert user: professional career in BCI context ($n = 13$). Age in years

No.	Gender	Age	Profession/role	Level of BCI experience
1	Female	29	Researcher	BCI expert
2	Female	35	Researcher	BCI expert
3	Female	25	Researcher	BCI expert
4	Female	25	Researcher	BCI expert
5	Female	24	Student	Naïve user
6	Male	30	Pilot	BCI expert
7	Male	32	Researcher	BCI expert
8	Male	28	Pilot	Naïve user
9	Male	28	Pilot	Naïve user
10	Male	25	Pilot	Naïve user
11	Male	32	Pilot	Naïve user
12	Male	56	Pilot	Naïve user
13	Male	26	Pilot	Naïve user
14	Male	32	Developer	Technical BCI expert
15	Male	31	Researcher	BCI expert
16	Male	44	Developer	Technical BCI expert
17	Male	50	Researcher	BCI expert
18	Male	26	Student	Naïve user
19	Male	44	Researcher	BCI expert
20	Male	45	Researcher	BCI expert
21	Male	48	Researcher	BCI expert
22	Male	55	Gamer	Naïve user
23	Male	31	Gamer	Naïve user
24	Male	30	Gamer	Naïve user

All interviews were audio-recorded and subsequently transcribed verbatim. The data was fully anonymized. We used the scientific software MAXQDA (version 2018) for content analysis according to Mayring [31, 32]. First, the analysis was done in German due to hermeneutic accuracy (e.g., metaphors, idioms). Later, it was translated into English for publication. We iteratively coded the interviews identifying general topics of interest and lines of argumentation. Using this method, three main categories emerged: (a) BCI experience from a healthy user perspective, (b) ethical evaluation, and (c) expectations and fears towards non-medical BCI use. The coding process was supported by an independent second researcher. Furthermore, the results were continuously validated in an interdisciplinary research team [32].

9.3 Results

What kind of experiences were associated with the use of BCIs? How was this unique way of human-machine interaction described? And what kind of ethical implications arose? Figure 9.1 presents the comprehensive category system. In the following, the results will be described in detail.

MC1: BCIs from a healthy user perspective
      C1: Ambivalent feelings
      C2: Uniqueness of the experience
      C3: Success as a decisive evaluation factor
      C4: Effective BCI use as a learning process
      C5: Human Factors in BCI use
                    C5.1: Cognitive exhaustion due to high concentration level
                    C5.2: Pressure to relax results in body tension
                    C5.3: Dry electrodes cause pressure and headache
                    C5.4: EEG-cap and -application
                    C5.5: BCI optics and miniaturization of technology
                    C5.6: The essential role of individualized BCI use
MC2: Ethical Evaluation
      C6: Control of action
      C7: Autonomy: Is the man or machine acting?
      C8: The question of responsibility
                    C8.1: Full responsibility
                    C8.2: Responsibility, only if use was successful
                    C8.3: No responsibility
                    C8.4: Objectivity through statistical measurement
                    C8.5: The question of responsibility as a social negotiation process
      C9: Privacy: Can BCIs read thoughts?
      C10: Updating humans? BCIs and the idea of man
                    C10.1: Are BCI users cyborgs?
                    C10.2: BCIs and human enhancement
MC3: Expectations and fears towards BCIs for healthy users
      C11: High expectations and Informed Consent
      C12: Legal regulations and inclusion of ethicists
      C13: Fears towards BCI technology
                    C13.1: Pro or contra BCIs? - An ambivalent evaluation process
                    C13.2: Risk of BCI-misuse
                    C13.3: Unfair advantages and fair access to neurotechnologies
      C14: A glimpse into the future of BCIs
                    C14.1: The future vision of a "Digital Companion"
                    C14.2: Active or passive BCI? - Future applications

Fig. 9.1 The qualitative category system; [*MC* main category; *C* category]

9.3.1   BCIs from a Healthy User Perspective

Our interviewees described BCIs as a tool or button - and a new way of communication. Comparisons with science fiction expressed the futuristic nature of this neurotechnology eliciting a high level of fascination and curiosity. After using BCIs for the first time, a user said: "It was a really unique experience […]. I've never experienced something like this again" (No. 6, l. 232–233).

Furthermore, an objectification of the human brain through BCI use became evident. For example, thoughts have been considered as "data" (No. 19, l. 105), "signals" (No. 7, l. 498), or "pattern" (No. 19, l. 211), sometimes even attributing human skills to the computer. On the one hand, the use of statistical methods implied that the success of BCI-modulated actions can be measured objectively. On the other hand, it became clear that BCI technology is still in its early stages of development.

Overall, we identified a strong desire for BCI use in our sample. Success with BCIs was highly dependent on individual motivation and training. The time-consuming application of EEG and the need to practice were regarded as the main hurdles for everyday use. Also, a high number of participants expressed being cognitively exhausted after the experience. In addition, the optical appearance of BCIs was perceived as unfavorable, especially the EEG cap and the application of gel for the electrodes created discomfort.

9.3.2 Ethical Evaluation of BCIs

In our sample, a high level of ambivalence regarding control and responsibility towards BCIs was evident. About half of the interviewees felt control and assumed responsibility over the use of the technology; the other half negated this statement completely. This feeling was highly dependent on successful BCI use and personal background knowledge. Due to their strong involvement in BCI technology, the expert users were aware of the current technical potential and level of reliability. The more they succeeded in accomplishing intended actions using BCIs, the more they accepted personal responsibility for it.

Furthermore, we asked our participants whether BCIs would lead to a change in the idea of man. The results reveal that individual self-determination is perceived as crucial. The original idea of man would be preserved as long as there are no social constraints in using BCIs for cognitive enhancement and the transformation of human capacities. However, participants with expert knowledge emphasized that this perception could change in the future:

> Right now […] it's completely inappropriate to drill a hole in someone's head unless there is a strong medical need for it, but people might start to see it differently over time. So maybe for example, they find that this can be very helpful for neurofeedback. That if you put a little chip in your head, you could learn to relax much more effectively […]. And people start to say, well, relaxation is health, I mean, if you can reduce stress, people might live longer (No. 17, l. 686–693).

Also, the metaphor of cyborgs was very often employed to describe BCI users with implanted electrodes.

9.3.3 Expectations and Fears Towards BCIs for Healthy Users

Our interviewees expressed a strong desire for regulation by law and highly supported the introduction of a BCI user license. An expert BCI user emphasized the learning process:

> You would have to get a license for BCIs just as much […] as for a normal car - by learning to operate it. If you can, you get your license and if you can't, you just don't get it (No. 10, l. 306–308).

In addition, a high level of unrealistic expectations regarding BCI technology became evident among naïve BCI users. Before using BCIs for the first time, the term "brain-computer interface" was highly linked to reading thoughts. Afterwards, however, the risk of mind reading was perceived as unlikely at the current stage of technology. Also, privacy issues and concerns about BCI misuse were reported: "You can use BCIs for good things and you can use it for not so good things and sometimes the situation is not black and white" (No. 24, l. 166–167).

The future potential of BCI technology was evaluated as extremely high, given an increased individualization of mental strategies and the continuously rapid technological development. Here, a possible change in society was described by expert users:

> Once enhancement by implants or BCIs in general will come […]. If so, then yes, then healthy people will start to use them like they use drugs, like they do use doping and sports (No. 15, l. 217–219).

Additionally, our participants agreed that in the context of healthy users passive BCIs are most promising and technological devices will be miniaturized in the future.

9.4   Discussion

Our results show that BCIs represent the fusion of man and machine in a unique way. Almost all participants highlighted being puzzled and fascinated by this new form of human-machine interaction. However, unrealistic expectations towards BCIs and the limited degree of reliability easily led to frustration. This ambivalence concerning BCIs was evident among almost all study participants. Informed consent is a central instrument for providing reliable information to study participants and ensuring respect for autonomy [33, 34]. Our findings reveal that it is not only important for patients, but also for healthy BCI users.

Furthermore, the question of motivation is crucial. Why should healthy people prefer active BCIs (and weeks of training) when they are able to do the same action manually within seconds using other technology? Also, the importance of individualized mental strategies became clear. Only a few publications regarding this topic exist so far [23, 35]. Therefore, further research in this field is strongly needed.

In addition, our results show that BCI-modulated actions were perceived as a mixture between man and machine. A clear and unanimous attribution of action and responsibility was no longer given. EEG technology is still prone to errors and requires a high level of personal expertise. Moreover, about 10% of users are not able to control a BCI successfully - a phenomenon called *BCI illiteracy* [36–38]. The complexity of BCI technology leads to a responsibility gap [3, 39, 40]. Thus, it's not surprising that a strong desire for regulations and laws regarding BCI use was evident in our sample. These findings are in accordance with existing literature where the claim for an active involvement of ethicists is also present [4, 22, 38, 41].

Here, the protection of the mental sphere and the preservation of free will are dominant factors [3, 41, 42]. We emphasize that the current legal situation stands in strong contrast to people's demands. To our knowledge, no regulatory framework for healthy users of intelligent neurotechnologies exists in Europe so far.

In the following, the methodological limitations of this work will be outlined. First of all, a high level of technology affinity was present in our selected group of interviewees. Most had technical professions and a high level of enthusiasm for technology. It cannot be ruled out that this selection bias influenced the evaluation in a positive way. Another challenge was the highly variable personal background knowledge and the level of BCI training they had. Eleven interviewees had tested BCIs only once. Also, some participants had experienced BCIs several months before. Others, however, were very experienced with BCIs and well-informed about the current technological state. This factor had a big influence on the individual attribution of responsibility for BCI-modulated actions. In addition, interviews were conducted in German and English making translation effects possible.

Finally, we would like to highlight the numerous ethical, legal, and social implications associated with non-medical BCIs. This paper aims to inform the general public about possible chances and risks of this neurotechnology. Only fully involved citizens and policymakers can make reliable and socially acceptable decisions.

Acknowledgments Work on this paper was funded by the Federal Ministry of Education and Research (BMBF) in Germany (INTERFACES, 01GP1622A). Financial support for J. R. Schmid from the Hanns-Seidel-Stiftung (BMBF funding) is gratefully acknowledged.

References

1. Graimann B, Allison B, Pfurtscheller G. Brain–computer interfaces: a gentle introduction. In: Graimann B, Pfurtscheller G, Allison B, editors. Brain-computer interfaces: revolutionizing human-computer interaction. Berlin: Springer; 2010. p. 1–27.
2. Shih JJ, Krusienski DJ, Wolpaw JR. Brain-computer interfaces in medicine. In: Mayo Clinic proceedings, vol. 87. Amsterdam: Elsevier; 2012. p. 268–79.
3. Steinert S, Bublitz C, Jox R, Friedrich O. Doing things with thoughts: brain-computer interfaces and disembodied agency. Philos Technol. 2018;32:457–82.
4. Ramadan RA, Vasilakos AV. Brain computer interface: control signals review. Neurocomputing. 2017;223:26–44.
5. Höhne J, Schreuder M, Blankertz B, Tangermann M. A novel 9-class auditory ERP paradigm driving a predictive text entry system. Front Neurosci. 2011;5:99.
6. Wolpaw J, Wolpaw EW. Brain-computer interfaces: principles and practice. New York: OUP; 2012.
7. Zander TO, Kothe C. Towards passive brain–computer interfaces: applying brain–computer interface technology to human–machine systems in general. J Neural Eng. 2011;8(2):025005.
8. Dobkin BH. Brain–computer interface technology as a tool to augment plasticity and outcomes for neurological rehabilitation. J Physiol. 2007;579(3):637–42.
9. Donati AR, Shokur S, Morya E, Campos DS, Moioli RC, Gitti CM, Augusto PB, Tripodi S, Pires CG, Pereira GA. Long-term training with a brain-machine interface-based gait protocol induces partial neurological recovery in paraplegic patients. Sci Rep. 2016;6:30383.

10. Chaudhary U, Birbaumer N, Ramos-Murguialday A. Brain–computer interfaces in the completely locked-in state and chronic stroke. In: Progress in brain research. Amsterdam: Elsevier; 2016. p. 131–61.
11. Buch E, Weber C, Cohen LG, Braun C, Dimyan MA, Ard T, Mellinger J, Caria A, Soekadar S, Fourkas A. Think to move: a neuromagnetic brain-computer interface (BCI) system for chronic stroke. Stroke. 2008;39(3):910–7.
12. Van Erp JBF, Lotte F, Tangermann M. Brain-computer interfaces: beyond medical applications. Comput IEEE Comput Soc. 2012;45(4):26–34.
13. Stawicki P, Gembler F, Volosyak I. Driving a semiautonomous mobile robotic car controlled by an SSVEP-based BCI. Comput Intell Neurosci. 2016;2016:1–14.
14. Fricke T. Flight control with large time delays and reduced sensory feedback. Munich: Technische Universität München; 2017.
15. Bos DP-O, Reuderink B, van de Laar B, Gürkök H, Mühl C, Poel M, Nijholt A, Heylen D. Brain-computer interfacing and games. In: Brain-computer interfaces. Berlin: Springer; 2010. p. 149–78.
16. Nijholt A. BCI for games: a 'state of the art' survey. In: International conference on entertainment computing. Berlin: Springer; 2008.
17. Gürkök H, Hakvoort G, Poel M. Evaluating user experience in a selection based brain-computer interface game a comparative study. In: International conference on entertainment computing. Berlin: Springer; 2011.
18. Blankertz B, Tangermann M, Vidaurre C, Fazli S, Sannelli C, Haufe S, Maeder C, Ramsey L, Sturm I, Curio G, Müller K-R. The Berlin brain–computer interface: non-medical uses of BCI technology. Front Neurosci. 2010;4:198.
19. Abney K, Lin P, Mehlman M. Military neuroenhancement and risk assessment. In: Neurotechnology in National Security and Defense. Boca Raton: CRC Press; 2014. p. 264–73.
20. Eaton ML, Illes J. Commercializing cognitive neurotechnology—the ethical terrain. Nat Biotechnol. 2007;25(4):393–7.
21. Kögel J, Schmid JR, Jox RJ, Friedrich O. Using brain-computer interfaces: a scoping review of studies employing social research methods. BMC Med Ethics. 2019;20(1):18.
22. Ahn M, Lee M, Choi J, Jun S. A review of brain-computer interface games and an opinion survey from researchers, developers and users. Sensors. 2014;14(8):14601–33.
23. Vuckovic A, Osuagwu BA. Using a motor imagery questionnaire to estimate the performance of a brain–computer interface based on object oriented motor imagery. Clin Neurophysiol. 2013;124(8):1586–95.
24. van de Laar B, Bos DP-O, Reuderink B, Poel M, Nijholt A. How much control is enough? Influence of unreliable input on user experience. IEEE Trans Cybernet. 2013;43(6):1584–92.
25. Nijboer F, Clausen J, Allison BZ, Haselager P. The Asilomar survey: stakeholders' opinions on ethical issues related to brain-computer interfacing. Neuroethics. 2013;6(3):541–78.
26. Schicktanz S, Amelung T, Rieger JW. Qualitative assessment of patients' attitudes and expectations toward BCIs and implications for future technology development. Front Syst Neurosci. 2015;9:64.
27. Palinkas LA, Horwitz SM, Green CA, Wisdom JP, Duan N, Hoagwood K. Purposeful sampling for qualitative data collection and analysis in mixed method implementation research. Adm Policy Ment Health Ment Health Serv Res. 2015;42(5):533–44.
28. Helfferich C. Leitfaden- und Experteninterviews. In: Baur N, Blasius J, editors. Handbuch Methoden der empirischen Sozialforschung. Wiesbaden: Springer VS; 2014. p. 559–74.
29. Helfferich C. Die Qualität qualitativer Daten. Berlin: Springer; 2011.
30. Döring N, Bortz J. Forschungsmethoden und Evaluation. Berlin: Springer; 2016.
31. Mayring P. Qualitative Inhaltsanalyse. Grundlagen und Techniken. Weinheim: Beltz; 2015.
32. Mayring P. Einführung in die qualitative Sozialforschung. Weinheim: Beltz; 2016.
33. McCullagh P, Lightbody G, Zygierewicz J, Kernohan WG. Ethical challenges associated with the development and deployment of brain computer interface technology. Neuroethics. 2014;7(2):109–22.

34. Mulvenna M, Lightbody G, Thomson E, McCullagh P, Ware M, Martin S. Realistic expectations with brain computer interfaces. J Assist Technol. 2012;6(4):233–44.
35. Friedrich EV, Neuper C, Scherer R. Whatever works: a systematic user-centered training protocol to optimize brain-computer interfacing individually. PLoS One. 2013;8(9):e76214.
36. Allison BZ, Neuper C. Could anyone use a BCI? In: Brain-computer interfaces. Berlin: Springer; 2010. p. 35–54.
37. Guger C, Daban S, Sellers E, Holzner C, Krausz G, Carabalona R, Gramatica F, Edlinger G. How many people are able to control a P300-based brain–computer interface (BCI)? Neurosci Lett. 2009;462(1):94–8.
38. Carmichael C, Carmichael P. BNCI systems as a potential assistive technology: ethical issues and participatory research in the BrainAble project. Disabil Rehabil Assist Technol. 2014;9(1):41–7.
39. Kellmeyer P, Cochrane T, Müller O, Mitchell C, Ball T, Fins JJ, Biller-Andorno N. The effects of closed-loop medical devices on the autonomy and accountability of persons and systems. Camb Q Healthc Ethics. 2016;25(4):623–33.
40. Haselager P. Did I do that? Brain–computer interfacing and the sense of agency. Mind Mach. 2013;23(3):405–18.
41. Van Erp J, Lotte F, Tangermann M. Brain-computer interfaces: beyond medical applications. Computer. 2012;45(4):26–34.
42. Ienca M, Haselager P. Hacking the brain: brain–computer interfacing technology and the ethics of neurosecurity. Ethics Inf Technol. 2016;18(2):117–29.

Diffusion on Both Ends: Legal Protection and Criminalisation in Neurotechnological Uncertainty

10

Susanne Beck

Contents

Abstract

The development of neurotechnology and AI will provide enormous challenges for the legal system. Traditional criminal law is only partly adequate in regulating the development of neurotechnology and AI. The difficulties in assessing offender and victim or the relevant interests to be protected prove that criminalising in such contexts of new technologies with unclear responsibilities and unforeseeable developments is inadequate or even unjust. Therefore, a new balance of interests has to be found. This includes new ideas for (legal) obligations of patients using neurotechnology combined with AI, as this could potentially endanger others. Additionally, it will be important to reassess the border between the physical and mental sphere and re-evaluate legal protection of the latter—the importance of one's identity and privacy has to be taken into account in the legal sphere more than it is done today. Also, we should reconsider the categories of "information", "data" and "personal data" when it comes to neurotechnology. In times of Big Data, of AI combining the different information and deducing more

S. Beck (✉)
Institute for Criminal Law and Criminology, Leibniz University Hanover, Hanover, Germany
e-mail: susanne.beck@jura.uni-hannover.de

© Springer Nature Switzerland AG 2021
O. Friedrich et al. (eds.), *Clinical Neurotechnology meets Artificial Intelligence*,
Advances in Neuroethics, https://doi.org/10.1007/978-3-030-64590-8_10

information and, in many cases, even the identity of the person, these borders are difficult to uphold. The specific re-balancing of all interests involved has to be done by the legislator.

10.1 Introduction

Neurotechnology, in combination with AI, will bring humans and machines even closer together than existing technologies already do. These traditionally diametrically opposed entities—humans on one side, machines on the other—are beginning to merge [1, 2]. This merger is not only apparent in that machines are becoming more humanlike and humans are starting to resemble machines in different ways; it is also meant literally, as machines are implanted into human bodies or human bodies are directly linked to machines, which are controlled not externally, but internally through processes of the brain—or do they control the processes of the brain? Investments supporting research in this area show the importance of these developments and potential social prospects [3, 4]. Thus, it is to be expected that developments will continue and affect different areas of human life.

10.2 Problem Analysis

These new technological developments have, as shown by the many contributions here, several ethical and legal consequences and require detailed considerations by these normative disciplines.

Before discussing these problems and questions in more depth, it is important to stress that this new technology also has potential advantages [5–7]; restricting the focus on the negative consequences could lead to an imbalanced evaluation. This chapter focuses on specific legal aspects and thus, it will not be possible to analyse all potential advantages in detail, but it should be pointed out that neurotechnology will probably allow previously untreatable diseases to be treated in new and better ways, as well as enhancing some human capacities [8, 9]. AI will enable machines to combine enormous amounts of information and make extremely fast decisions in connection with many other machines and platforms that can improve the quality of these decisions (faster, better informed, etc.).

Still, the specifics of neurotechnology, which creates a direct link between humans and machines, raises even more questions than other new technologies—especially in combination with the unpredictable and, in many ways, uncontrollable technology of AI.

The challenges of this new technology are mostly based on the fact that integrating machines into the human body (brain) permits access to a person's brain and thoughts, provides different and new ways to communicate via the brain (BCI) [10] and interact via brains; which could (intentionally or not) alter the brain's processes either right away or in the form of long-term effects or cause unintentional external movements [5]. These effects have various implications that we will discuss later. In the context of AI, these specific problems stem from the "autonomy" of machines and their development [11], caused by deep learning [12], which connects numerous programs with machines and "transfer[s] of some part of decisions" onto the machines [11, 13]. The specific decision is unpredictable beforehand for both the programmer and the user and the entire process is, in many ways, uncontrollable and unclear, which leads to many new challenges. Thus, the combination of these new technologies also leads to a combination of their problems.

Some ethical aspects that are being discussed with regard to neurotechnology are privacy, consent, identity, agency and equality [3]. The technological access to a person's brain leads to concerns about the information received, a person's privacy and security against misuse and cyber-attacks [14]. Neurotechnology does of course bear risks for the patient on whom it is being tested. As is the case for every new technology and every new medical treatment, the risk is unknown and unpredictable; the probabilities of the realisation of specific risks cannot be predicted. This might hinder fully informed consent, although the patient can agree to the uncertainty of the risks. Tampering with brain processes and thus changing behaviour or altering emotions, memories or thoughts could cause identity problems and raise questions about who the agent of specific actions is, especially if the person whose brain is altered can still be regarded as agent in the traditional normative sense. And, of course, the focus also has to be directed towards equality—not just because the research is expensive and this kind of treatment will only be available for a part of the world's population, but also because if this technology is used to enhance one's capacities it will give that person advantages. These might not always be regarded as justified and thus lead to inadequate inequalities.

Although ethical concerns will not be discussed in detail here, they are, as we will show, also relevant for the legal debate—to show potential conflicts, possible solutions and the different interests of the parties involved. Moreover, of course, these concerns raise many more legal questions than can be addressed here, be it questions concerning data protection, admission of such machines, product safety or the conditions of research and usage of these machines in general [15]. More specifically, these last points are connected to legally relevant risk-benefit assessments, meaning that the admission of these types of products or medical therapies typically requires said assessment [16]. As already mentioned in the context of informed consent, it is difficult to deal with the uncertainty of the risks involved with this technology and the ambiguity of the potential consequences of a technology that can have an enormous impact on individuals, violate their rights, endanger their interests, etc. Therefore, a risk-benefit assessment in the context of neurotechnology combined with AI is especially difficult, which makes many legal aspects problematic [17–19].

Nonetheless, our focus here will be put on criminal law, including the perspective of the potential offender as well as the potential victim. This means that we will focus on the legal protection of the patient on one side and the protection of third persons from the patient on the other side. In this context, we will also analyse the criminalisation of these patients as well as of other parties involved. As will be shown, this technology might lead to diffusion on both ends, on the end of the victim as well as the offender. What we mean by diffusion and how to evaluate this development will be analysed in detail.

10.3 Legal Protection by Criminalisation

One function of the law is to protect members of society and defend their rights and interests, especially those who are not fully able to protect themselves (due to a power imbalance, age, mental or physical restraints, missing knowledge, skills or other reasons) ([20] marginal no. 5).

Criminal law is intended to focus on protecting the most relevant and important interests ([20] marginal no. 9; [21, 22], p. 40). In the context of neurotechnology, it is necessary to prevent the patient from using inadequate, risky technology and treatments, which could lead to violations of his body without his consent. It might also be necessary to prevent inadequate self-enhancement—this will have to be discussed. It is also necessary in order to protect others from the negative side effects, as far as this is possible by criminal law.

10.3.1 Protecting Patients

Neurotechnological research and treatment might permit "reading" someone's mind, communicate via BCIs or from brain to brain, collect data about the patient's brain processes and connect these with other patients' data or other data from this patient. Using AI, this also means that the data will be combined and used as a training basis for self-learning systems. This could mean a strong invasion into the privacy of the patient [3, 7, 15, 23]—it is hard to imagine more sensitive data than one's brain processes, thoughts and emotions. Thus, it is crucial to secure strong protection of these interests and rights. The data protection regime in the EU, the new General Data Protection Regulation (GDPR) of 2018 [24], can be regarded as quite strict. In consequence, it is under discussion whether these strict rules may hinder the development of AI and thus should be loosened [25]. The necessity of data protection is especially problematic, given the possibility to personalise data (which might be anonymous to start with), as well as the legal restriction of humans being addressed by machine decisions. While we are not able to discuss the data protection regime at this point—our main focus will be put on criminal law—it is important to note that the balance between these interests should be re-evaluated by society, not by the interpreters of the law; thus, at the moment, the strictness of the data protection regime has to be respected by legal practice. In the context of sensitive data, such as that necessary for neurotechnological development, the importance of strict rules is self-evident.

Additionally, neurotechnology traditionally requires surgery or other invasive measures in the bodily sphere of the patient [15, 18, 26]. Therefore, as with every other medical treatment, the patient has to be protected from treatment executed against his will, from wrongful or risky treatment or from being used for research that does not benefit him but only others/the doctor [27].

In the context of neurotechnology and AI this means dealing with the fact that the treatment is not just new and unknown, but also very risky and, in combination with AI, in many ways unforeseeable and uncontrollable. Therefore, this means one has to discuss, first of all, if and how informed consent is possible in cases where the patient cannot provide conclusive information [23, 26, 28]. Of course, it would be problematic to forbid such research and treatment, because in general, the patient has a human right to receive health treatment and is allowed to undergo risks and dangers according to his will. Every medical treatment involves risks and medical research regularly requires dangerous testing and participants willing to take these risks. Thus, risks themselves do not necessitate forbidding research and development as such [29].

To ensure that the patient is not misused as an object in on-going research, it is especially important, in the context of new and untested medical technologies, to remain open and honest and give the patient full disclosure on the status of the research, the known risks and all potential risks, which may still be undefined at that moment [30–32]. It is also important to note that in medical law—as well as in medical ethics—there are restrictions concerning research on participants that exhibit no medical necessity (e.g. the participant is either healthy or the treatment is unsuitable for his illness) [33]. This is also the case for research on patients that are unable to consent themselves, either because they are minors or because of an existing mental illness. This is especially important in the context of neurotechnology as it could be used to cure some mental illnesses, but the research might be problematic—here, in the light of the enormous risks, one would have to discuss if and when it could be acceptable for representatives to consent on behalf of the patient. This might be the case if it becomes clear that the patient could profit from the treatment.

Obviously, caution must be heeded in the risk-benefit analysis of AI-supported neurotechnology as well as in cases of highly vulnerable patients. Conflicts could also arise if the research and treatment is actually necessary to establish the capability to communicate or to decide for themselves [34]. In this case, the patient is unable to decide on participating in the risky treatment or research, conversely it is precisely this measure that could actually enable him to make this decision. In this case, one could let the parents or other representatives make the decision under specific conditions for the patient even if the treatment is still in the research stage, as it is clearly in the interest of the patient to be able to communicate or make decisions [29, 33].

Neurotechnology, in some cases, can be regarded as "enhancement", signifying that brain functions are improved without a medical need [32]; the patient might just want to become more intelligent, enhance some emotions or sensitive capacities [8, 9, 35]. For our considerations, it is not relevant if or when this is actually possible; the question is if the legislator can or should forbid such undertakings. Here, one could regard any legal limitations of such procedures as serving to protect the "patient" from himself. Paternalistic restrictions are problematic, even if they partly protect the greater interests of society, such as the above-mentioned equality [36]. This problem does not differ vastly from the general debate on enhancement in modern medicine, though [37]. Again, the unknown and uncontrollable risks enlarge the problems, but the general direction of the arguments—if and under which conditions paternalism is acceptable—stays the same.

One problem that is quite specific is that the intrusion into the brain structure and even changing brain functions might be possible without invasive methods [38]. This means that the bodily sphere remains intact, but the brain functions will be influenced by technology. Although the effect will probably be similar with regard to personality, character, capabilities, etc. and the potential negative side effects of the stimulation of the brain will exceed those following the surgical procedure, it is not at all clear whether non-invasive stimulation of the brain is actually forbidden by criminal law at the moment. German criminal law does forbid a violation of the body, but it does not incriminate a violation of the mental state [39]. It has to be discussed if influencing brain functions can be considered a violation of the bodily sphere; but even if such interpretation of the law is possible, this cannot cover the meaning of the crime. The real violation of such stimulation does not lie in changing the body, but in the potential changes in one's identity, thoughts or emotions, in the possible intrusion into one's privacy, and therefore in changes in the mental dimension instead of the physical one [40].

As we have seen in the ethical debate, identity is one of the main concerns in the context of neurotechnology/AI [41–43].

Thus, it seems necessary to at least discuss new legislation that provides protection for the mental state, as well [44, 45]. This does not mean that one has to be in favour of such law, especially criminal law forbidding any interference with another's mental state—it could be difficult to differentiate between socially adequate and inadequate influences on others' mental states, as we all influence each other on a daily basis, and inacceptable interferences; meaning that the machine cannot be the relevant difference. It will also be problematic to use criminal law as the "sharpest weapon of the state" or "ultima ratio" ([20] marginal no. 15; [46–50]). However, that does not mean that the imbalance should not be discussed further.

Protecting third parties? It is also necessary to protect others, in the sense of individuals that might be harmed by the patient after and because of the neurotechnological treatment, but also protecting society at large from unwanted neurotechnology. However, the latter might not be protectable by criminal law. Although arguments are often in favour of paternalistic criminal laws in that they also protect society from inadequate developments, imbalances or injustices, slippery slopes [51] (e.g. of values), etc., this is not entirely plausible. Criminal law should always be the ultima ratio and its usage should also be proportionate: it should be generally restricted to protect vital interests and it should also be functional. In the case of vague social interests that might not even be shared by all members of society, criminal law is mostly inadequate, but more importantly, it does not work to protect such interests in this way—criminal laws only have an impact if they represent the real existing social values of the majority.

Nevertheless, even the protection of individuals from the patient could be difficult. Neurotechnological treatment could lead to personality changes in the patient and, in some cases, to aggression or other behaviour that violates the rights and interests of others. To prevent this from happening, though, one might have to forbid use of the technology altogether; this, again, seems unproportionate regarding the possible advantages for patient, though. Therefore, one has to discuss the question of responsibility of the patient—can he still be held responsible and also, which actions and which conditions can be regarded as (criminally) responsible [52].

In general, the existing criteria for criminal responsibility can be used to determine if the patient should be sanctioned after neurotechnological treatment or not.

There will be cases in which the patient cannot be regarded as being capable of understanding the law or acting accordingly at the precise moment of the criminal action, which is required by German criminal law for being held responsible ([20] marginal no. 605, 617). Still, it seems problematic to not hold the patient responsible if he knew beforehand that his incapability might be one of the consequences of neurotechnological treatment. This is currently under discussion in Germany concerning the brain-pace-maker, in that the patient often knows about the subsequent personality changes before he activates the machine, and might even know that he is likely to commit crimes as this "other" person [41, 52]. Therefore, one could argue that he might not be responsible for the action, but for activating the machine knowing about the probability of committing a crime as a result (so-called actio libera in causa) [52]. This, again, would have the problematic consequence that he could in some cases, not be allowed to activate the machine although it might be the only way to relieve the symptoms of a severe illness [52].

These "black or white" solutions, meaning either not to allow the treatment at all or not holding the patient responsible, do seem inadequate. A compromise could be to make the patient responsible for his omissions instead of his actions [52]. For example, if the patient knows that he tends to shoplift in a specific shop after he activates the brain-pace-maker, he could be obliged to inform the people working in the shop to look after him; if he is known to become aggressive it could be necessary for him to protect others from himself, either by being accompanied or even by ensuring that he is locked in [52]. This way, potential victims are protected quite adequately while the patient does not have to give up treatment. These considerations can be applied to each neurotechnology, even AI-supported neurotechnology, that has the potential to alter the patient's personality; although one has to bear in mind that AI is harder to predict so that the potential risks for others might not be as clear and easy to prevent. In this case, one has to re-evaluate the balance between potential risks for others and the patient's interests. The unknown risks might only be acceptable for even more severe illnesses; the obligation to protect society from oneself might be higher if the patient does not know all the potential actions that he may be capable of if receiving specific treatment, etc. [52].

Again, the difference between "treatment" and "enhancement" becomes important at this point [52]. Society is more likely to accept the motive of receiving neurotechnological treatment for one's benefit and to treat a severe illness, even taking into account the risks for others as well as advantages over others. The reason for this is that "others" are only potentially violated by this person because he can overtake them.

Another aspect that seems to be new and interesting is criminal responsibility when communicating via BCI or communicating and collaborating with another human via his/her brain. It is possible that the patient could use this communication or collaboration to commit crimes—if the technology is just used as a means then there are no relevant new aspects compared to other means in which the usage is, of course, criminalised. And, of course, if one of the people connected via brain uses the other person to commit crimes, then the used person is, in many cases, not responsible; although, like the patient mentioned above, under certain circumstances, the used person could be held responsible for allowing the other one to use them. But if the crime is committed under collaboration between two humans connected via brains and it cannot be proven who thought and who acted, it is difficult to assess who should be criminalised. This is the expression of a general diffusion of responsibility in the context of artificial intelligence. Wherever new forms of collaboration between humans and humans and machines emerge, it challenges the traditional ways of assigning responsibility [11, 53]. In the case of two humans connected via brain it might be possible that, regarding "in dubio pro reo"[1] [20], no one could be criminalised for a certain crime under criminal law.

10.3.2 Diffusions

Let us now analyse the diffusions we find in these developments. In order to understand the difficulties of the legal system, it is necessary to regard the diffusions of traditional legal categories—with these diffusions come injustices, impracticalities and thus the need for adjustments, as we will see.

The next categories that seem to diffuse in the context of neurotechnology are "victim" and "offender". While traditional crimes mostly permit a clear separation between the two, it does not seem as easy here: the patient can, in some ways, be regarded as the victim and in other ways as the potential offender of the specific occurrence [41]. He is a potential victim regarding his privacy, his bodily sphere and his mental state—and this not at some point in this life (before the offence) but at the same time, in the same instance, diffusing the categories of being either offender or victim of a specific offence. But he is also a potential offender, not just against himself, thus paternalistic laws have to protect him from himself, but also possibly protect society if the development is unwanted, and—esp. important—protect single individuals, if his behaviour changes significantly after the treatment and he becomes aggressive [54, 55] or violates others' rights. Society or "the others" can be regarded as potential victims of the patient or, in this case, of him becoming aggressive, but also in the case of enhancement in that his actions might lead to

---

[1] According to the principle in dubio pro reo, the conviction of the defendant for a criminal offense is only possible if the court is satisfied that he has committed it. If the judge has any doubts after having taken the evidence, he must assume that this is more favorable for the accused.

unwanted developments and social pressure on others [3, 7, 36, 56]. But they can also be offenders—the doctor who might undertake the research or treatment, the parties that might be interested in knowing the information from the brain processes of the patient or at least using the data to develop technologies, or for other purposes, etc. [57, 58].

The diffusion of the categories offender and victim is important for the usage and relevance of criminal law in this context. Traditional criminal law—in theory—intends to manifest trust in norms in society, it only punishes clearly offensive and socially inadequate behaviour and is based on generally accepted values [21, 22]. This also means that it is very clear who is the offender and who is the victim. The described diffusion crushes the legitimacy of criminalising someone who is not clearly the offender or to protect someone who is not clearly only a victim. Although the diffusion might not be as strong in the context of neurotechnology, it is at least strong enough so that criminalising loses some of its legitimacy and might not be as useful to reach the traditional aims of criminal law.

Another diffusion can be found when looking at the interests and rights that are to be protected. Criminal law, up until now, has set clear borders between the body and the mind, protecting the bodily sphere, yet remaining mostly absent from the protection of the mental state ([39], p. 102; [40, 41, 45]). This has different reasons; the interference with others' mental state is often unavoidable and socially adequate and differentiating between these actions and unwanted violations is difficult. Only some violations of specific rights, such as the victim's reputation or informational self-determination (privacy), are protected by criminal law.

But now, with the growing importance of neurotechnology and knowledge about the mental state of the patient and how to influence it, this juxtaposition can be questioned. Interfering with the brain's functions clearly leads to changes in the mental state, starting from a physical level. The relevance of this interference does not lie in the bodily sphere but in changing thoughts, emotions, maybe even the personality of the patient [15]; this unacceptability does not change significantly based on whether it is done internally via surgery or externally. Thus, it is necessary to question this focus on the body, the dualism that the law is based on, and to rethink the interests that should be protected by criminal law [40, 42, 43, 45].

Similarly, the interests of data and privacy could become unclear in the context of neurotechnology. Images of brain functions are in itself not personal, per se, not data—if depersonalised and made anonymous and used for research as such ([3, 7, 59], p. 394)—in the traditional sense and just producing such anonymous images in itself can hardly be regarded as a clear violation of the patient's privacy. It is different when we consider the interpretation of these images as well as their usage in the context of AI in that, by comparing images of numerous persons as well as using additional information might lead to being able to learn to "read someone's mind"—even if the data has been made anonymous beforehand, to understand and to collect the deepest thoughts and emotions of persons [15]. Thus, these developments lead to obscurity about what is protectable data, what the nucleus of our concept of privacy is and how it can and should be protected in the context of Big Data and AI.

10.3.3 Some Reflections on Future Debates

This overview of the diffusions shows that traditional criminal law is only partly adequate in regulating the development of neurotechnology and AI. The difficulties in assessing offender and victim or the relevant interests to be protected prove that criminalising in such contexts of new technologies with unclear responsibilities and unforeseeable developments is often not just functional, but also inadequate or even unjust.

A new balance of interests has to be found, for example, for the obligations of patients using neurotechnology combined with AI, which could potentially endanger others. The unforeseeability of AI leads to difficult considerations on how many risks others have to bear or which obligations the patient has to have to avoid these risks. Here it could be important that the legislator provides some legal certainty, that he regulates the obligations at least partly. Admittedly, it is difficult to foresee all potential situations and regulate all potential obligations in advance; but it might be an adequate start to differentiate between treatment and enhancement. In terms of treatment, in my opinion, it should not be forbidden, but the patient should have an obligation to avoid potential risks as far as he is aware of them. In terms of enhancement, one should consider restricting it legally if it becomes apparent that inadequate risks for others arise because of personality changes among patients. One could also consider that the person in this case does stay responsible for actions after the treatment; whether civil responsibility for damages can be considered as sufficient or if it is necessary to induce criminal responsibility will have to be discussed, as well.

Additionally, it will be important to reassess the border between the physical and mental sphere and re-evaluate legal protection of the latter. Although, as mentioned, there are many difficulties to consider when regulating violations of the mental sphere, the importance of one's identity—and privacy—has to be taken into account. It will be unavoidable to find solutions that protect these spheres better than is the case today. In my opinion, this should be done by upholding the categories "body" and "mind", as the relevant violations will differ. The exact determination of actions that should be forbidden will have to be decided by the democratically legitimated legislator.

Similarly, I think we should reconsider the categories of "information", "data" and "personal data" when it comes to neurotechnology. The images produced and the information received might not be personal data in the traditional sense, if it has been made anonymous for research. But in times of Big Data, of AI combining the different information and deducing more information and, in many cases, even the identity of the person, these borders are difficult to uphold. This does not mean that a strict data protection regime should be applied to these developments in general—instead, one should discuss how we can protect our information and data on the one side, but still allow the further development of neurotechnology and AI on the other. Again, this re-balancing has to be done by the legislator, but it is important to keep the different relevant interests in mind when reassessing the situation.

In general, the development of neurotechnology and AI will provide enormous challenges for the legal system. Each of these topics in itself shatters traditional concepts and categories such as identity and responsibility. In combination, these difficulties become even more prominent. This does not mean that the advantages of these technologies should be sacrificed, though. Instead, the meeting of neurotechnology and AI could be an adequate moment to reassess current legal concepts and categories and to adjust outdated ones.

References

1. Kurzweil R. Spiritual machines: the merging of man and machine. Futurist. 1999;33(9):16.
2. Warwick K. The merging of humans and machines. In: Neurotechnology, electronics, and informatics. Cham: Springer; 2015. p. 79–89.
3. Yuste R, Goering S, Bi G, Carmena JM, Carter A, Fins JJ, et al. Four ethical priorities for neurotechnologies and AI. Nat News. 2017;551(7679):159.
4. Report: Newly published market research report projects neurotechnology market will reach $13.3B by 2022. 2018. https://www.globenewswire.com/news-release/2018/04/23/1485369/0/en/Newly-Published-Market-Research-Report-Projects-Neurotechnology-Market-Will-Reach-13-3B-by-2022.html.
5. Müller O, Rotter S. Neurotechnology: current developments and ethical issues. Front Syst Neurosci. 2017;11:93.
6. Lynch Z. Neurotechnology and society (2010–2060). Ann N Y Acad Sci. 2004;1013(1):229–33.
7. Kellmeyer P. Big brain data: on the responsible use of brain data from clinical and consumer-directed neurotechnological devices. Neuroethics 2018:1–16.
8. Farah MJ. Emerging ethical issues in neuroscience. Nat Neurosci. 2002;5(11):1123.
9. Farah MJ, Illes J, Cook-Deegan R, Gardner H, Kandel E, King P, et al. Neurocognitive enhancement: what can we do and what should we do? Nat Rev Neurosci. 2004;5(5):421.
10. Schalk G, McFarland DJ, Hinterberger T, Birbaumer N, Wolpaw JR. BCI2000: a general-purpose brain-computer interface (BCI) system. IEEE Trans Biomed Eng. 2004;51(6):1034–43.
11. Beck S. The problem of ascribing legal responsibility in the case of robotics. AI Soc. 2016;31(4):473–81.
12. LeCun Y, Bengio Y, Hinton G. Deep learning. Nature. 2015;521(7553):436.
13. Clausen J. Neurotechnologien interdisziplinär: Anthropologische und ethische Überlegungen. 2011. https://leibniz-institut.de/archiv/clausen_20_03_11.pdf.
14. Ienca M, Haselager P. Hacking the brain: brain–computer interfacing technology and the ethics of neurosecurity. Ethics Inf Technol. 2016;18(2):117–29.
15. Ienca M, Andorno R. Towards new human rights in the age of neuroscience and neurotechnology. Life Sci Soc Policy. 2017;13(1):5.
16. Taupitz J, Schreiber M. Zulässige Maßnahmen im Rahmen von nicht-interventionellen Studien. Pharma-Recht. 2015;37(12):573–83.
17. Giordano J. Toward an operational neuroethical risk analysis and mitigation paradigm for emerging neuroscience and technology (neuroS/T). Exp Neurol. 2017;287:492–5.
18. Farah MJ. An ethics toolbox for neurotechnology. Neuron. 2015;86(1):34–7.
19. Horstkötter D. The ethics of novel neurotechnologies: focus on research ethics and on moral values. AJOB Neurosci. 2016;7(2):123–5.
20. Wessels J, Beulke W, Satzger H. Strafrecht Allgemeiner Teil, 40. ed. Heidelberg: 2015.
21. Bock D. 2. Kapitel: Funktion des Rechts; Funktion des Strafrechts: Strafzwecke. Strafrecht Allgemeiner Teil. Berlin: Springer; 2018, p. 61–75.
22. Jakobs G. Strafrecht, Allgemeiner Teil: Die Grundlagen und die Zurechnungslehre. Berlin: Walter de Gruyter; 2011.

23. Roelfsema PR, Denys D, Klink PC. Mind reading and writing: the future of neurotechnology. Trends Cogn Sci. 2018;22(7):598–610.
24. Bublitz JC. Privacy concerns in brain–computer interfaces. AJOB Neurosci. 2019;10(1):30–2.
25. Mertens P. Die Datenschutz-Grundverordnung–eine kritische Sicht. Wirtschaftsinformatik & Management. 2019:1–12.
26. Giordano J. A preparatory neuroethical approach to assessing developments in neurotechnology. AMA J Ethics. 2015;17(1):56–61.
27. Vlek RJ, Steines D, Szibbo D, Kübler A, Schneider M-J, Haselager P, et al. Ethical issues in brain–computer interface research, development, and dissemination. J Neurol Phys Ther. 2012;36(2):94–9.
28. Farisco M, Evers K. Neurotechnology and direct brain communication: new insights and responsibilities concerning speechless but communicative subjects. Abingdon: Routledge; 2016.
29. Clausen J. Ethische Aspekte der neurowissenschaftlichen Forschung. Handbuch Ethik und Recht der Forschung am Menschen. New York: Springer; 2014. p. 457–64.
30. Clausen J. Ethical brain stimulation–neuroethics of deep brain stimulation in research and clinical practice. Eur J Neurosci. 2010;32(7):1152–62.
31. Kirchhof G. Ärztliches Handeln zwischen Heilung, Forschung und Erneuerung Schutz des Patienten durch eine modifizierte Typologie ärztlicher Eingriffe. MedR Medizinrecht. 2007;25(3):147–52.
32. Suhr K. Der medizinisch nicht indizierte Eingriff zur kognitiven Leistungssteigerung aus rechtlicher Sicht. Berlin: Springer-Verlag; 2015.
33. Prütting J. Rechtliche Aspekte der Tiefen Hirnstimulation: Heilbehandlung, Forschung, Neuroenhancement. Berlin: Springer-Verlag; 2013.
34. Haselager P, Vlek R, Hill J, Nijboer F. A note on ethical aspects of BCI. Neural Netw. 2009;22(9):1352–7.
35. Giordano J. The human prospect (s) of neuroscience and neurotechnology: domains of influence—and the necessity—and questions—of neuroethics. Hum Prospect. 2014;4(1):1–18.
36. Wikler D. Paternalism in the age of cognitive enhancement: do civil liberties presuppose roughly equal mental ability? In: Savulescu J, Bostrom N, editors. Human enhancement. Oxford: Oxford University Press; 2009. p. 341.
37. Beck S. Enhancement–die fehlende rechtliche Debatte einer gesellschaftlichen Entwicklung. MedR Medizinrecht. 2006;24(2):95–102.
38. Seitz F. Die Tiefe Hirnstimulation im Spiegel strafrechtlicher Schuld. Berlin: Springer; 2020.
39. Knauer F. Der Schutz der Psyche im Strafrecht. Tübingen: Mohr Siebeck; 2013.
40. Bublitz JC. Der (straf-) rechtliche Schutz der Psyche. Rechtswissenschaft. 2011;2(1):28–69.
41. Bublitz C, Merkel R. Guilty minds in washed brains? In: Vincent NA, editor. Neuroscience and legal responsibility. New York: Oxford University Press; 2013. p. 335–74.
42. Merkel R. Personale Identität und die Grenzen strafrechtlicher Zurechnung: Annäherung an ein unentdecktes Grundlagenproblem der Strafrechtsdogmatik. Juristenzeitung 1999:502–11.
43. Merkel R. Neuartige Eingriffe ins Gehirn–Verbesserung der mentalen condicio humana und strafrechtliche Grenzen. Zeitschrift für die gesamte Strafrechtswissenschaft. 2009;121(4):919–53.
44. Von AM. Cyborgs und Brainhacks: Der Schutz des technisierten Geistes. In: Albers M, Katsivelas I, editors. Recht & Netz. 1st ed. Baden-Baden: Nomos Verlagsgesellschaft mbH & Co. KG; 2018. p. 491–518.
45. Bublitz JC, Merkel R. Crimes against minds: on mental manipulations, harms and a human right to mental self-determination. Crim Law Philos. 2014;8(1):51–77.
46. Gärditz KF. Demokratizität des Strafrechts und Ultima Ratio-Grundsatz. JuristenZeitung. 2016;71(13):641–50.
47. Hoven E. Was macht Straftatbestände entbehrlich?–Plädoyer für eine Entrümpelung des StGB. Zeitschrift für die gesamte Strafrechtswissenschaft. 2017;129(2):334–48.
48. Jahn M, Brodowski D. Das Ultima Ratio-Prinzip als strafverfassungsrechtliche Vorgabe zur Frage der Entbehrlichkeit von Straftatbeständen. Zeitschrift für die gesamte Strafrechtswissenschaft. 2017;129(2):363–81.

49. Prittwitz C. Das Strafrecht: Ultima ratio, propria ratio oder schlicht strafrechtliche Prohibition? Zeitschrift für die gesamte Strafrechtswissenschaft. 2017;129(2):390–400.
50. Kindhäuser U. Straf-Recht und ultima-ratio-Prinzip. Zeitschrift für die gesamte Strafrechtswissenschaft. 2017;129(2):382–9.
51. Volokh E. The mechanisms of the slippery slope. Harv L Rev. 2002;116:1026.
52. Beck S. Neue Konstruktionsmöglichkeiten der actio libera in causa. ZIS. 2018;4:204–11.
53. Kellmeyer P, Cochrane T, Muller O, Mitchell C, Ball T, Fins JJ, et al. The effects of closed-loop medical devices on the autonomy and accountability of persons and systems. Camb Q Healthc Ethics. 2016;25(4):623–33.
54. Lim S-Y, O'Sullivan SS, Kotschet K, Gallagher DA, Lacey C, Lawrence AD, et al. Dopamine dysregulation syndrome, impulse control disorders and punding after deep brain stimulation surgery for Parkinson's disease. J Clin Neurosci. 2009;16(9):1148–52.
55. Sensi M, Eleopra R, Cavallo MA, Sette E, Milani P, Quatrale R, et al. Explosive-aggressive behavior related to bilateral subthalamic stimulation. Parkinsonism Relat Disord. 2004;10(4):247–51.
56. Lachenmeier M. Sozialer Druck im Kontext von Human Enhancement: Autonomie und Freiwilligkeit bei der Anwendung von pharmakologischen, neurotechnologischen und gentechnischen Massnahmen zur Steigerung der Leistungsfähigkeit des Individuums. Basel: Helbing Lichtenhahn; 2017.
57. Eaton ML, Illes J. Commercializing cognitive neurotechnology—the ethical terrain. Nat Biotechnol. 2007;25(4):393.
58. Olson S. Brain scans raise privacy concerns. Washington, DC: American Association for the Advancement of Science; 2005.
59. Schaar P. Leitplanken für die digitale Gesellschaft. Digitalisierung im Spannungsfeld von Politik, Wirtschaft, Wissenschaft und Recht. New York: Springer; 2018. p. 387–95.

Data and Consent Issues with Neural Recording Devices

11

Stephen Rainey, Kevin McGillivray, Tyr Fothergill, Hannah Maslen, Bernd Stahl, and Christoph Bublitz

Contents

Abstract

Research-driven technology developments in the neurosciences present interesting and potentially complicated issues concerning data in general and, more specifically, brain data. The data that is produced from neural recordings is unlike names and addresses in that it may be produced involuntarily, and it can be processed and reprocessed for different aims. Its similarity with names, addresses,

S. Rainey (✉)
Oxford Uehiro Centre for Practical Ethics, University of Oxford, Oxford, UK
e-mail: stephen.rainey@philosophy.ox.ac.uk

K. McGillivray
University of Oslo, Oslo, Norway

T. Fothergill · B. Stahl
De Montfort University, Leicester, UK

H. Maslen
University of Oxford, Oxford, UK

C. Bublitz
Faculty of Law, University of Hamburg, Hamburg, Germany

© Springer Nature Switzerland AG 2021
O. Friedrich et al. (eds.), *Clinical Neurotechnology meets Artificial Intelligence*,
Advances in Neuroethics, https://doi.org/10.1007/978-3-030-64590-8_11

etc. is that it can be used to identify persons. The collection, retention, processing, storage and destruction of brain data are of high ethical importance. In terms
of policy, as one strand of a broader fabric of measures to cope with this, we can
ask: is current data protection regulation adequate in dealing with emerging data
concerns that relate to consumer neurotechnology and consent?

11.1   Questions and Scope

This chapter provides an overview of some of the sorts of question that might arise
from a data protection perspective as they relate to consent in the use of brain-
computer interfaces. This ought to aid in promoting a better understanding of these
technologies from a potential consumer perspective, especially by discussing various possibilities for recording and processing brain signals, which are likely to
appear in consumer technologies. Highlighting these common areas of concern for
consumers will provide product developers with insight into the context that their
products will be used in. This is important for consumers to whom the products will
be marketed, and to the regulatory frameworks that will most likely constrain the
functions of these products. Moreover, brain-signal recordings that are generated
from direct to consumer (DTC) devices provide data of an indeterminate type (is it
personal data, medical data?). Discussion of these issues is required to bring neuro-
data issues to the attention of policymakers.

At least two relevant points concerning data and consent emerge in cases of neural recordings:

1. Recording: While a user can easily consent to using a device to record data for
   specific purposes, recording involuntary data may occur without user consent.
2. Processing: What does processing data entail and does this have ramifications for
   consent?

Informed consent has a long heritage as a sine qua non for human research. It
forms part of good clinical practice and is enshrined as such in the European
Commission Directive 2001/20/EC, which describes the implementation of good
clinical practice in the conduct of clinical trials on medicinal products for human use:

Informed consent is the decision, which must be written, dated and signed, to
take part in a clinical trial, taken freely after being duly informed of its nature, significance, implications and risks and appropriately documented, by any person
capable of giving consent or, where the person is not capable of giving consent, by
his or her legal representative; if the person concerned is unable to write, oral consent in the presence of at least one witness may be given in exceptional cases, as
provided for in national legislation.[1]

The broader policy position on informed consent appears in the Declaration of
Helsinki, referencing the necessity of ethical considerations:

---

[1] The full directive can be viewed at https://ec.europa.eu/health/human-use/clinical-trials/direc-
tive_en [accessed March 2019].

The design and performance of each research study involving human subjects must be clearly described and justified in a research protocol. The protocol should contain a statement of the ethical considerations involved and should indicate how the principles in this Declaration have been addressed.[2]

Before going on to treat these issues in terms of consent, some limitations need to be established. First, we will limit ourselves to examining data in terms of the European General Data Protection Regulation (GDPR) [1].[3] This limitation is advisable owing to its scope, the authors' familiarity with it and the availability of space that precludes a more widespread treatment. Second, we will not engage in a detailed legal analysis of contract law and consumer law where 'consumer BCI devices' are mentioned. One reason for this is lack of space. Another is the future-facing nature of the technology under scrutiny. We wish to maintain a higher level of analysis so that we can anticipate ethical concerns that may be on or beyond the horizon. By operating at this level, we will be able to maximise the scope and applicability of our analysis.

11.2   How Neurotechnologies Work

Neurotechnologies typically work by reading, recording and processing brain signals. In research contexts, the process of recording permits various degrees of invasiveness from scalp-based electroencephalogram (EEG), brain surface electrode (electrocorticography, or ECOG) arrays to intracortical probes. Consumer contexts tend to use scalp-based type EEG approaches, for obvious reasons. These can be used to operate different types of devices including wheelchairs, prosthetic limbs, drones or software programs. They can be put to use for rehabilitation, or sometimes as means of neuro-optimisation and enhancement by providing users with feedback on their neural activity [2–5]. The therapeutic potential of these approaches is tantalising, in allowing individual, brain-based ways to overcome problems in mobility, affective disorders, cognitive impairment, or in fine-tuning brain processes according to will. But the idea of 'mind'-controlled devices is more widely seen as an interesting and exciting mode of engagement with technologies for consumer applications or creative pursuits [6–12]. Does the nature of neural recording, processing and modulating pose a potential risk in terms of consent and how the data derived from brains is used?

There is increasing public understanding about issues surrounding consenting to online data collection and use [13, 14]. Yet, there still remains much complacency and ongoing misuse of these data [15]. Many major technology companies have been implicated in over-reaching with their data collection activities. The consequences for the public have included reputational damage to political systems following microtargeted voter swaying campaigns, as in the 2016 UK Brexit referendum [16, 17]. These are issues that may potentially have far-reaching consequences and implications.

---

[2]World Medical Association, 2000, paragraph 22, describes the context for informed consent http://www.wma.net/en/30publications/10policies/b3/ [accessed March 2019].

[3]Subsequent references to specific parts of GDPR legislation refer to the English language version of the regulation.

Social media companies that enable this kind of activity have reportedly seen user activity change. While user numbers have fallen in general, greater awareness of privacy-impacting conditions on such platforms has resulted in changing attitudes towards participation, as well as changing attitudes towards privacy itself [18–20]. Political concern and media attention have focussed on slow responses to dealing with data problems [21, 22]. While a lot of data obtained by social media companies and others is given voluntarily (if not advisedly), more data than is sometimes realised is obtained by obfuscated means. A wide variety of data can be collected without the express knowledge of the user, and thus only with a very dubious sense of their 'consent' (See for instance [23]). Inferences can be drawn about user activity in general, based upon data derived from what kind of web history a user has, for example, or from their location data as recorded while using the internet from a mobile phone. This provides a clear parallel with brain-signal recording that is worth drawing out.

11.3   Neural-Signal Recording

Neural data has multifarious uses in research contexts to indicate memory content, motor and speech intention, mood and educational aptitude, among other things. As such, this represents data that could easily be framed as sensitive, personally identifying and revelatory about a person. This is made all the more acute by the growing market in consumer neurotechnology. Although neural data is mostly collected for medical and neuro-science research, the recent increase in digital health technologies beyond research and medical facilities to non-invasive and readily accessible consumer-grade applications raises issues of privacy and informed consent. If we are unprepared, or underprepared, in dealing with online data collection and use, we ought to be bracing for the potential risks inherent with neural data collection, as this is of a more intimate and malleable form than web history or location data.

If a neural device is intended as a neural-controller for a piece of hardware or software, relevant biosignals can be extracted from neural recordings in order to trigger, control and optimise that device or application. However, other information could be derived from the same recordings of those signals by means of subsequent reprocessing and interpreted in ways which present ethical concerns. This is perhaps especially the case given the rate at which recording density is increasing, the greater understanding of how inter-neuron communication affects information processing and the likely increased future role for machine learning in neural data analysis [24].

In terms of the GDPR, individuals whose data is being collected, held or processed are referred to as 'data subjects'. Everyone is at some point a data subject, and may be a data subject in a variety of different ways for different contexts. Given the fact that neural data may be collected, held and processed in various ways, users of neurotechnologies are data subjects in this way. Importantly, especially for regulation as will be discussed in the next section, neural data can be sensitive enough to render a data subject identifiable. Perhaps more importantly from a practical point of view, these signals could be *taken to* identify a data subject, regardless of whether or not they *actually* do. As such, they could represent a significant issue for a data subject in terms of their otherwise private neural state becoming an apparently open resource from which to characterise them somehow.

We will now take some key concepts from the GDPR and then relate them back to the neural-recording context in order to tease out the implications for consumers and policies.

11.4   The GDPR

Concerns around ethical issues arising from exponential use of brain data such as personality invasion, mental integrity and breach of personhood draw attention to the need to examine legal responses that seek to prevent adverse effects for data subjects [25]. Recital 15 of the GDPR provides neutrality for technology that specifically caters to the protection of natural persons irrespective of the technologies used. This provision aims to reduce the risk of circumvention of the law by recognising potential issues of emerging technologies and their impacts on data protection and seeks to balance the competing interests of privacy and technology [26].

As mentioned above, the GDPR provides data protection measures centred upon persons conceived of as data subjects. One aspect of this includes restrictions and other conditions on the processing of personal data. To assess the role of personal data processing in the context of brain-signal recordings for neurotechnologies, we need to ascertain whether such recordings constitute personal data, and if so, is processing involved.

11.5   Is Brain Data Personal Data?

Personal data in the GDPR is any data that can identify a natural person.

GDPR Art 4 (1) 'personal data' means any information relating to an identified or identifiable natural person ('data subject'); an identifiable natural person is one who can be identified, directly or indirectly, in particular by reference to an identifier such as a name, an identification number, location data, an online identifier or to one or more factors specific to the physical, physiological, genetic, mental, economic, cultural or social identity of that natural person;

Moreover, the European Court of Justice in the case of Breyer vs Germany ruled that the *possibility* of identification is sufficient to consider some data personal data [27]. Readily available neural-recording techniques, like consumer-grade EEG or fMRI, can possibly be used to identify persons [28–31]. Examining the recordings from individuals' brains and using various signal-processing techniques can identify subjects with high accuracy in certain circumstances. It is not implausible to think that these techniques will improve. This ought to prompt discussion over brain reading and mental privacy [32].

Further questions might be raised as to whether certain types of neural recordings might be considered medical data, over and above personal data. According to GDPR Art 4 (15),

… 'data concerning health' means personal data related to the physical or mental health of a natural person, including the provision of health care services, which reveal information about his or her health status;

It is certainly the case that neural recordings can serve to indicate diseases such as epilepsy [33]. They can also indicate affective states, including traumatic memory [34]. This permits the possibility of identifying data subjects and their physical or mental health. For now, this analysis will not stray too far into the question of medical data, but will instead just note that it serves to illustrate how one recording may have implications wider than the purposes it is ultimately made for. Neural signals do not need to be recorded for the purpose of identifying data subjects or for identifying physical or mental disease. That information may nonetheless be derived at some stage, through subsequent processing, or through association with other data from other sources, perhaps. The nature of processing neural recordings is of central importance for this reason.

11.6 Is Personal Data Processed in Neurotechnological Devices?

Using the information mentioned, it seems likely that neural recordings in themselves or in combination with other factors are personal data in terms of the GDPR, which brings us to the question of how these personal data are processed. According to GDPR Art 4 (2), it seems clear that neurotechnological devices or systems that use brain-signal recordings also process personal data. In this context, 'processing'

…means any operation or set of operations which is performed on personal data or on sets of personal data, whether or not by automated means, such as collection, recording, organisation, structuring, storage, adaptation or alteration, retrieval, consultation, use, disclosure by transmission, dissemination or otherwise making available, alignment or combination, restriction, erasure or destruction;

Almost any use of the data, even storage, will be considered 'processing'; collecting it in the first place, assessing it, storing it, sharing it, analysing it, technologically processing it. The key is therefore identifying the data subject and the possibility of linking that data to the subject. If brain data is personal data, and it is processed in particular ways in neurotechnological systems, then it may be characterised as biometric data under the GDPR.

GDPR Art 4 (14) 'biometric data' means personal data resulting from specific technical processing relating to the physical, physiological or behavioural characteristics of a natural person, which allow or confirm the unique identification of that natural person, such as facial images or dactyloscopic data;

As has been referenced already, even EEG recordings can be used to identify subjects. But more to the point, various data that is prohibited by GDPR Art 9 may be processed. This article prohibits the processing of

...biometric data for the purpose of uniquely identifying a natural person, data concerning health or data concerning a natural person's sex life or sexual orientation shall be prohibited.

Neural recordings can be used to predict age [35]. Sex and age can also be predicted where recordings are made for reasons other than making that prediction, in sleep research for instance [36]. Brain recordings have been used to investigate differences among different genders and sexualities, such as the neural difference between homosexual men and heterosexual men and women [37]. Even though these predictions were revealed to be inaccurate in specific cases, the ethical issues arising from the identifiability of persons on that basis remain. A *mis*identification can be just as problematic as an accurate identification, especially where sensitive personal characteristics are involved.

Organisational or social measures can be used to minimise the risks for data subjects in contexts of complex data collection and use. By adding additional layers between collection, retention, processing and the retrieval of data, the chances of misuse or accidental leaks can be minimised. For example, let's imagine Company *A* stores its customers' personal data. The personal data is sent to service provider *B* where a lookup table is created and the data are anonymised. The data are then sent to service provider *C* and put to use in some application (e.g. product functionality, debugging). Service provider *C* will not have access (or rights) to the lookup table held at *B*. Service provider *B* will have a very high level of security and anonymous data-handling expertise. A setup such as this is what can appear in contexts of medical patient data research.

There are risks to this approach, as well. For instance, a data breach at service provider *B* would allow *C* to identify the data subjects. In terms of the European Commission sourced advice for compliance with GDPR, from the advice of Working Party 29,[4] the table held at *B* would need to be deleted (possibly also the underlying data) for the data to be considered properly anonymous. However, especially in the context of medical research, that raises ethical issues regarding incidental findings and, perhaps, the validity of the research, not least in terms of reproducibility.

Perhaps more directly relevant to the context of consumer BCIs using neural recordings, rather than by way of analogy as in medical research and patient data, is the idea of purpose specification. For most data-processing activities, a specific purpose must be set out prior to the use of that personal data. For example, if a company wishes to process my data for the purpose of delivering a service (e.g. updating some software I have purchased from them), they cannot automatically re-use my information for marketing purposes. In scientific research, there is scope for such further use—repurposing of the data—owing to the nature of scientific research as being open-ended.

In scientific research, the scientist(s) may not know how the research will end at the onset (see GDPR Recital 33 and Article 89). But in terms of consumers'

---

[4] This was a working party set up to provide advice while the GDPR was being developed. Its work can be seen at https://ec.europa.eu/justice/article-29/documentation/opinion-recommendation/index_en.htm [accessed March 2019].

personal data this is not so clear. We need to expand on the recording of brain signals in order to follow up on the potential for repurposing data and how it may raise conceptual and ethical issues relating to consent.

11.7   Recording Brain Signals and the GDPR

In terms of consenting to having brain signals recorded in the first place, questions arise about consenting to potentially unknown outcomes. The question here amounts to how a *specific consent* can be made into a *general collection* with a *wide scope* for repurposing. The stakes are high in terms of the nature of the data as personal or biometric.

Let's imagine a specific scenario where a consumer agrees to the terms of use for a neuro-controlled robotic arm:

Ada likes technology and eagerly purchases the robotic arm. It is controlled via an EEG type cap, with a few electrodes. These electrodes are positioned on the cap with the stated aim of recording brain signals associated with motor cortex activity. Through some training, Ada will be able to control the robotic arm by imagining moving her own limbs. These imagined movements will realise neural activity that the software for her device will come to recognise as control commands for the robotic arm.

It seems clear that Ada agrees to the use of her neural recordings as a control parameter for the robotic arm. The possibility of identifying persons from simple EEG recordings has been alluded to above, so already we see this, at least implicitly, as an agreement to the use of personal data for this purpose. Specifically, she has agreed to motor activity being recorded—that is how the device is stated to work. It is worth noting that some suggest that systems like these are more likely to use facial muscle activity rather than neural activity [38]. But at any rate, the brain is increasingly understood as an open system with distributed functionality. Given the nature of EEG recordings as scalp-based, identifying specific brain areas and specific signals is not simple. Recording brain signals at this level is quite general. Simply due to the physical fact of the EEG cap's distance from the sites of motor neural activity and the impedance of the brain and skull themselves, this may be insufficient to limit the recordings made to purely motor signals.

Recordings of brain signals can be processed in order to create information about specific neural activity, which can be used to infer behavioural, dispositional or other personal data. The scope of the original brain recordings' interpretation is therefore of central importance. The nature of a brain-signal recording as a source of information is open to modification given differing techniques used for processing. It is clear that *existing* signals can be transformed so as to reveal more, or different, information than what they represented at the time of recording. This is what was meant above by distinguishing among specific consent, general collection and a wide scope for repurposing. This makes the processing of data particularly salient.

The nature of data retention, storage and destruction is most pertinent in the case of processing. If 'raw' recordings from Ada's device are retained, they are apt to be reprocessed in ways that were not necessarily consented to. This could amount to a serious repurposing of data, especially in the light of the GDPR. The data might be processed so as to become biometric data, as per Art 4 (14) mentioned above. These could be used to identify Ada or to infer things about her physical and mental state, her age, gender, sexual orientation and so on. This may be done well or poorly. Furthermore, where AI is involved, the possible scope for repurposing data may not be understood or anticipated fully ahead of time. With this, the issues here are parallel to the ethical issues attending Big Data in general.

Data in general is being collected from a wide variety of devices, in a range of contexts, at an incredible rate (e.g. sensors, internet use, mobile phones). Because of its nature it can't be said to be 'stored' in any conventional sense of the term. For one thing, it is dynamic and it is constantly being updated. It is for these reasons that it is not conventionally accessible, and so is not available without some amount of processing. Big Data provides seemingly endless possibilities for predicting, refining and reconceptualising various domains. These include how financial markets operate, the analysis of social realities and how research is carried out [39]. Beyond data mining [40], Big Data offers a dynamic and huge resource that makes big promises for its users, but which raise ethical concerns [41].

Especially in terms of personal and health data, issues arise concerning the ownership, monetisation and privacy of data. The type of processing that Big Data undergoes, which is necessary to access any of its promised insights, is algorithmic. It is not necessarily related to any particular dataset, but may be a set of sets or a set of sub-sets. The kinds of patterns recognised in data points by algorithms do not preclude the crossing of boundaries among data points. This is the point of algorithmic processing in one sense, as it 'sees' patterns in huge amounts of data that a human would not be able to see.

In this context, the connection between brain data and Big Data can be seen. Big Data can be deployed to answer questions that were not asked at the time of data collection. This does not make data a set of information specific to a research question and a methodology, as might conventionally be the case. Instead, a 'discovery science' attitude can be adopted by researchers in terms of the data itself. It can also lead to a sense in which researchers expect the data to answer questions that have not been thought of yet, leading to a data-driven approach replacing a knowledge-driven one [42, 43]. This leads some in health research to

> …hope we do not fall into the trap of believing that any new information technology is worth using in health research regardless of the ethical issues in performing the research or the larger implications. [44]

AI operating on existing data can, perhaps unpredictably, produce new information or predictions. However, the nature of a neurotechnology system may require that past data be used. Even if this is done only for debugging or optimising the system as a whole, it represents a secondary use of data and may have problematic dimensions if data is re-purposed or identifies specific data subjects.

Systems may require data in order to function optimally. In other words, data collected at one time for a specific purpose may be grouped with other data and processed in order to provide diagnostic data for the system overall. In the context of consumer devices, individuals' recordings may serve to optimise a system-as-product in a general sense. Ada's robotic arm may be one of a million units sold. Each new firmware update may rely on the use of each user's data being processed in some very general sense. The status of the user's data in such a context is open to question: Is it research data? It seems it can't be anonymous, in any full-blooded sense, as a neural-controlled system adapts to the user as much as they are trained to use that system [45].

At least at some level, the kind of optimisation data, containing lots of processed data from lots of users, must be fed back specifically for each user. This would be how the general processing would lead to specific device optimisation. We must then ask, in the light of the GDPR: How is this curated? The data subject must be identifiable for the optimisation feedback, but the optimisation process of the grouped data seems to rely upon personal data being generally processed. How do we get the flour back from the dough once the loaf is baked? How could it be destroyed, in case Ada grows wary of all this and does not want it anymore? In terms of the GDPR, this amounts to a question about the implementation of the so-called 'right to be forgotten'. When a system-as-product relies on a confluence of multiple users' data collected over time, this seems (at least) very difficult. Central to these questions is the role of consent. With such varying, wide-ranging and apparently open possibilities for the eventual fate of data derived from neural recordings, the possibility of consenting to those recordings is complex.

11.8 Consent

Current data protection regulation strongly emphasises consent. Consent is one of the six legal bases for processing personal data under Article 6 (1) of the GDPR. Processing personal data is generally prohibited, unless the data subject has consented to it or if it is expressly allowed by law.

> GDPR Art 4 (11) 'consent' of the data subject means any freely given, specific, informed and unambiguous indication of the data subject's wishes by which he or she, by a statement or by a clear affirmative action, signifies agreement to the processing of personal data relating to him or her;[5]

Specific information must be provided to data subjects where personal data are collected under Article 13(2), including the right to withdraw consent and information regarding automated decision-making or profiling. The GDPR is an improvement from earlier data privacy regimes in that it further promotes and protects the interests of data subjects, including matters of consent, portability and erasure. Current data protection regulation has a strong campaign in favour of consent. As such, wherever data appear, it ought in principle to be identifiable as related to a data subject and, to some degree, be under their control. For instance, if a company holds data on a data subject, that subject ought to be able to get a full rundown of the data held, to rectify inaccurate personal data or insist upon its destruction.

---

[5] Basic requirements for consent are provided in Article 7 and further specified in Recital 32 of the GDPR.

BCIs, in principle, prompt specific issues regarding consent due to their focus on neural signals and the functions of the brain, and thus could have very intimate links to concepts of the self [46]. Given that some users of BCIs may be ill, specific considerations are needed in terms of the major decision-making they may be implicated in. Users of speech prostheses, for example, would require careful attention where end-of-life planning was at stake [47]. But even before these kinds of considerations become relevant, there is the question of the data's collection, storage and processing. These data-specific questions raise consent issues of their own, requiring practicable solutions [48].

Neural devices operate through the extraction of brain signals from neural data that may be produced involuntarily, difficult to individuate because of the recording technique involved and perhaps processed by AI. Depending on what kind of signal is required, different data may be extracted from one set of neural recordings. As has been suggested already, recording brain signals is general while processing brain signals is purpose-relative.

Consent represents an essential factor in how we might conceptualise data issues that may attend neurotechnologies in the near future. It is just one dimension of this conceptualisation, however, in a field growing in complexity, reach and import. How data, once collected, might be reprocessed later has been the focus throughout this chapter, and this hinges on consent. Broad consent, modelled on a bio-banking approach, may be too passive or too reliant on expert decision-making on behalf of data subjects [49]. Dynamic consent, however, relying on technologies to permit user changes to data use may be onerous [50]. Moreover, these may represent other issues in that data would have to be used and re-used in the very process of consent. This could be another issue of the GDPR in terms of data encryption, minimisation and destruction, for instance.

Given the scope for possible uses of brain data and its reprocessing and repurposing, one fear might be that consent itself comes to be seen as an impediment to innovation for this technology. This fear, in which consent comes to be seen as a problem rather than a safeguard, is already discussed in the context of Big Data [51]. Consent, at one very general level, is a means of recognising the autonomy of individuals with all the protections such recognition brings with it, such as non-domination, non-coercion and respect. Challenges to this ought not to be taken lightly. Neurotechnology developers and consumers in general ought to learn from the context of Big Data and avoid this unfortunate outcome.

Specific types of consent, such as those modelled on 'broad consent' as seen in bio-banking contexts, or a technology-driven dynamic consent approach, ought to be bolstered by an effort to encourage a general understanding of neurotechnologies. A better understanding of how neural-based devices work will allow informed decision-making on potential future implications for consenting to brain data collection. This suggests that no particular regime of consent will solve all possible issues arising. A culture of understanding must accompany any solution in order to establish informed attitudes towards technology. Developers of technologies ought to resist overstating or mis-stating the possibilities of their devices in order not to allow a sense of mystery to overcome clarity. This might prevent clear decisions on practical problems regarding data use that would otherwise present themselves.

11.9 Conclusion

It seems that, on the present reading of the GDPR, brain-reading devices present non-trivial consent issues for consumers and developers of the technology. At a conceptual level, they appear to offer only a very complex and convoluted possibility for consenting to their use. This, combined with the likely nature of the way in which the systems will work, makes them a potential challenge for current European data protection standards. Ethically, this is a problem for consumers and developers. This ethical problem could become a legal one if the user of a neurotechnological device were to find themselves unable to exercise their rights owing to the nature of the product they are using.

This has not been a legal analysis, but rather an analysis of the principle of consent and the idea of compliance with the GDPR. There appear to be difficulties, especially where AI processing is involved, in reconciling these factors for neuro-technological devices. This means that better consumer understanding of the stakes must be forthcoming, and that a general awareness of these issues ought to be given from the outset when designing a neuro-controlled device.

References

1. European Commission (2016) Regulation (EU) 2016/679 of the European Parliament and of the Council of 27 April 2016 on the protection of natural persons with regard to the processing of personal data and on the free movement of such data, and repealing Directive 95/46/EC (General Data Protection Regulation) (text with EEA relevance).
2. Young BM, Nigogosyan Z, Nair VA, et al. Case report: post-stroke interventional BCI rehabilitation in an individual with preexisting sensorineural disability. Front Neuroeng. 2014;7:18. https://doi.org/10.3389/fneng.2014.00018.
3. Sitaram R, Ros T, Stoeckel L, et al. Closed-loop brain training: the science of neurofeedback. Nat Rev Neurosci. 2017;18:86–100.
4. Roelfsema PR, Denys D, Klink PC. Mind reading and writing: the future of neurotechnology. Trends Cogn Sci. 2018;22:598.
5. Nuffield Council on Bioethics, editor. Novel neurotechnologies: intervening in the brain. London: Nuffield Council on Bioethics; 2013.
6. Pan P, Tan G, Wai P, Aung A. Evaluation of consumer-grade EEG headsets for BCI drone control. In: International researcher club conference on science, engineering and technology, 2017. p. 10–1.
7. Kapur A, Kapur S, Maes P. AlterEgo: a personalized wearable silent speech interface. In: 23rd international conference on intelligent user interfaces. ACM; 2018. p. 43–53.
8. Forrest C. Facebook planning brain-to-text interface so you can type with your thoughts. In: TechRepublic; 2017. https://www.techrepublic.com/article/facebook-planning-brain-to-text-interface-so-you-can-type-with-your-thoughts/. Accessed 14 Nov 2018.
9. Whyte C. Mind-reading headset lets you Google just with your thoughts. New Scientist. 2018.
10. Revell T. Mind-reading devices can now access your thoughts and dreams using AI. New Scientist. 2018.
11. Ings S. Pierre Huyghe at the serpentine—digital canvases and mind-reading machines. Financial Times. 2018.
12. Akst J. Decoding dreams. The Scientist Magazine. 2013.
13. Milne GR, Rohm AJ, Bahl S. Consumers' protection of online privacy and identity. J Consum Aff. 2004;38:217–32.

14. Boyd D, Hargittai E. Facebook privacy settings: who cares? First Monday. 2010. https://doi.org/10.5210/fm.v15i8.3086.
15. Cadwalladr C, Graham-Harrison E. The Cambridge analytica files. The Guardian. 2018. p. 6–7.
16. Howard PN, Kollanyi B. Bots, #Strongerin, and #Brexit: computational propaganda during the UK-EU referendum. Rochester: Social Science Research Network; 2016.
17. Del Vicario M, Zollo F, Caldarelli G, Scala A, Quattrociocchi W. Mapping social dynamics on Facebook: the Brexit debate. Soc Networks. 2017;50:6–16.
18. Teens, Social Media & Technology 2018 | Pew Research Center; 2018.
19. Lankton N, McKnight D, Tripp J. Understanding Facebook privacy behaviors: an integrated model. Faculty Research Day. 2018.
20. Tsay-Vogel M, Shanahan J, Signorielli N. Social media cultivating perceptions of privacy: a 5-year analysis of privacy attitudes and self-disclosure behaviors among Facebook users. New Media Soc. 2018;20:141–61.
21. Collins D, Efford C, Elliot J, et al. Disinformation and 'fake news'. London: The Digital, Culture, Media and Sport Committee; 2019. p. 111.
22. Wong JC. Facebook acknowledges concerns over Cambridge Analytica emerged earlier than reported. The Guardian. 2019.
23. Linus R. What every browser knows about you. In: Webkay; 2019. http://webkay.robinlinus.com/. Accessed 25 Mar 2019.
24. Stevenson IH, Kording KP. How advances in neural recording affect data analysis. Nat Neurosci. 2011;14:139–42.
25. Nijboer F, Clausen J, Allison BZ, Haselager P. The Asilomar survey: stakeholders' opinions on ethical issues related to brain-computer interfacing. Neuroethics. 2013;6:541–78.
26. European Commission. The GDPR: new opportunities. Brussels: New Obligations; 2018.
27. Borgesius FZ. The Breyer case of the court of justice of the European Union: IP addresses and the personal data definition. Eur Data Prot L Rev. 2017;3:130.
28. Bashar K. ECG and EEG based multimodal biometrics for human identification. In: 2018 IEEE international conference on systems, man, and cybernetics (SMC); 2018. p. 4345–50.
29. Abdullah MK, Subari KS, Loong JLC, Ahmad NN. Analysis of the EEG signal for a practical biometric system. World Acad Sci Eng Technol. 2010;4:5.
30. Gui Q, Jin Z, Xu W. Exploring EEG-based biometrics for user identification and authentication. In: Proceedings of the IEEE signal processing in medicine and biology (SPMB); 2014. p. 1–6.
31. Horien C, Shen X, Scheinost D, Constable RT. The individual functional connectome is unique and stable over months to years. NeuroImage. 2019;189:676. https://doi.org/10.1016/j.neuroimage.2019.02.002.
32. Mecacci G, Haselager P. Identifying criteria for the evaluation of the implications of brain reading for mental privacy. Sci Eng Ethics. 2019;25(2):443–61.
33. Staba RJ, Stead M, Worrell GA. Electrophysiological biomarkers of epilepsy. Neurotherapeutics. 2014;11:334–46.
34. Steinert S, Friedrich O. Wired emotions: ethical issues of affective brain–computer interfaces. Sci Eng Ethics. 2019;26:351. https://doi.org/10.1007/s11948-019-00087-2.
35. Al Zoubi O, Ki Wong C, Kuplicki RT, Yeh H, Mayeli A, Refai H, Paulus M, Bodurka J. Predicting age from brain EEG signals—a machine learning approach. Front Aging Neurosci. 2018;10:184. https://doi.org/10.3389/fnagi.2018.00184.
36. Carrier J, Land S, Buysse DJ, Kupfer DJ, Monk TH. The effects of age and gender on sleep EEG power spectral density in the middle years of life (ages 20 60 years old). Psychophysiology. 2001;38:232–42.
37. Alexander JE, Sufka KJ. Cerebral lateralization in homosexual males: a preliminary EEG investigation. Int J Psychophysiol. 1993;15:269–74.
38. Wexler A, Thibault R. Mind-reading or misleading? Assessing direct-to-consumer electroencephalography (EEG) devices marketed for wellness and their ethical and regulatory implications. J Cogn Enhanc. 2018;3:131. https://doi.org/10.1007/s41465-018-0091-2.

39. Mittelstadt BD, Floridi L. The ethics of big data: current and foreseeable issues in biomedical contexts. Sci Eng Ethics. 2016;22:303–41.
40. Fan W, Bifet A. Mining big data: current status, and forecast to the future. SIGKDD Explor Newsl. 2013;14:1–5.
41. Prinsloo P, Slade S. Big data, higher education and learning analytics: beyond justice, towards an ethics of care. In: Daniel BK, editor. Big data and learning analytics in higher education. Berlin: Springer International Publishing; 2017. p. 109–24.
42. Bollier D, Firestone CM. The promise and peril of big data. Washington, DC: Aspen Institute, Communications and Society Program; 2010.
43. Boyd D, Crawford K. Critical questions for big data. Inf Commun Soc. 2012;15:662–79.
44. Rothstein MA. Ethical issues in big data health research: currents in contemporary bioethics. J Law Med Ethics. 2015;43:425–9.
45. Vaughan TM, Heetderks WJ, Trejo LJ, Rymer WZ, Weinrich M, Moore MM, Kübler A, Dobkin BH, Birbaumer N, Donchin E. Brain-computer interface technology: a review of the second international meeting; 2003.
46. Klein E, Ojemann J. Informed consent in implantable BCI research: identification of research risks and recommendations for development of best practices. J Neural Eng. 2016;13:043001.
47. Glannon W. Brain-computer interfaces in end-of-life decision-making. Brain Comput Interfaces. 2016;3:133–9.
48. Kim H, Bell E, Kim J, Sitapati A, Ramsdell J, Farcas C, Friedman D, Feupe SF, Ohno-Machado L. iCONCUR: informed consent for clinical data and bio-sample use for research. J Am Med Inform Assoc. 2017;24:380–7.
49. Steinsbekk KS, Kåre Myskja B, Solberg B. Broad consent *versus* dynamic consent in biobank research: is passive participation an ethical problem? Eur J Hum Genet. 2013;21:897–902.
50. Kaye J, Whitley EA, Lund D, Morrison M, Teare H, Melham K. Dynamic consent: a patient interface for twenty-first century research networks. Eur J Hum Genet. 2015;23:141–6.
51. Kaye J. Abandoning informed consent: the case of genetic research in population collections. In: 267412; 2004. https://repository.library.georgetown.edu/handle/10822/503951. Accessed 20 Feb 2017.

Ethical Implications of Brain-Computer Interface and Artificial Intelligence in Medicine and Health Care

12

Ralf J. Jox

Contents

Abstract

Artificial intelligence and brain-computer interfaces are two novel technologies that have numerous potential areas of application in medicine. They raise, however, significant ethical implications that call for reflection and discussion before deciding about the use of these kinds of applications. In this chapter, I present some examples of these technologies, focusing first on the ethical implications of medical research on brain-computer interfaces. Using the example of a recent case of alleged scientific misconduct, I highlight the dangers inherent in this kind of research on clinical technology. Second, I focus on ethical issues in the clinical application of artificial intelligence and deep learning algorithms in medicine and highlight some risks and challenges for the patient–physician relationship, but more fundamentally also for the character of medicine.

R. J. Jox (✉)
Institute of Humanities in Medicine and Clinical Ethics Unit, Lausanne University Hospital and University of Lausanne, Lausanne, Switzerland
e-mail: ralf.jox@chuv.ch

© Springer Nature Switzerland AG 2021
O. Friedrich et al. (eds.), *Clinical Neurotechnology meets Artificial Intelligence*,
Advances in Neuroethics, https://doi.org/10.1007/978-3-030-64590-8_12

155

12.1   Introduction

The development of medicine in our days is profoundly driven by the application of ever new technologies. Currently, two classes of novel technologies are at the brink of shaping the medicine of the future: artificial intelligence (AI) and neurotechnologies that directly interact with the human brain, in particular brain-computer interface (BCI). The potential of these technologies is vast, but so are also their risks and ethical implications. In this chapter, I would like to draw attention to some of these ethical implications that BCI and AI have on the practice of medicine and health care today.

Before embarking on this task, I would like to make three important preliminary remarks: (1) AI and BCI are, of course, two distinct technologies with different ethical implications. However, invasive neurotechnologies, in particular BCI, increasingly incorporate AI and deep learning so that the former is not fully comprehensible without the latter. (2) The implications of these new technologies go far beyond ethical issues, they touch on legal, social, philosophical, economic, political, and other dimensions of human life. To be coherent with the focus of this book, however, I will focus on the *ethical* implications, in particular on research ethics and clinical ethics, while being aware that the ethical dimension overlaps with these other dimensions. (3) Given the breathtaking pace of these technological inventions and their broad potential uses I can only proceed in a cursory way in this chapter, selecting some of the most remarkable forms of technology use and some of the most salient ethical issues arising thereof.

12.2   The Importance of Neuropsychiatric Disorders for the Development of Health Care Technology

Disorders of the nervous system are among the most common diseases worldwide, and their frequency is rising due to global aging [1]. They contribute most significantly to the global burden of disability. Recent advances in early (even predictive) diagnosis and in disease-modifying therapies both result in even longer chronicity and a higher prevalence of such disorders as Alzheimer's disease, Parkinson's disease, or multiple sclerosis.

One of the characteristics of nervous system disorders is that they often have "negative" rather than "positive" symptoms, loss of function rather than unpleasant sensations like pain, nausea or dyspnea. Patients lose basic functions of their everyday life, such as walking, using their hands, practicing hygiene, speaking, or swallowing. The capacities that people tend to lose with nervous system disorders are commonly capacities that are highly valued in our society: cognition, communication, personal autonomy, responsibility, biographical life planning, and social interaction. This contrast explains much of the enormous impact that neuropsychiatric diseases have on individuals, families, and society. It is no surprise that dementia is tending to replace cancer as the most feared illness [2]. These fears create a very

Box 12.1: A Fictitious case example

Fred was a 58-year-old farmer from Bavaria, Germany, and a passionate amateur soccer player. One day he noted difficulties while handling the gear lever of his tractor and when opening the door of his house with his keys. Over the next weeks, both of his hands became more and more clumsy and weak. The diagnostic workup at his primary care physician was inconclusive, so he was sent to an orthopedic physician who presumed the diagnosis of a cervical spine injury and operated on him. The operation, however, did not change anything for the better and it took another 5 months until a neurologist finally established the diagnosis of amyotrophic lateral sclerosis. At that time, Fred already had some degree of paralysis in all four limbs. Shortly after, his speech became slurred and his family and friends suspected him to be an alcoholic, which is why he increasingly withdrew from social contacts and became very isolated. When he finally lost his ability to express himself verbally, he used a tablet computer to communicate, but even this became more and more cumbersome for him. One day, when he was surfing the Internet, he read about a new study on a device named brain-computer interface, which allows paralyzed persons to speak, move, and be autonomous again. He immediately seized this opportunity and registered for the study, dreaming of a new life despite his illness and disability…

strong emotional incentive and a societally compelling reason to conduct research on nervous system disorders.

One particular disease that is among the most feared (and that often nurtures wishes to hasten death [3]) is amyotrophic lateral sclerosis (ALS), also called motor neuron disease. This disorder is caused by a continuously progressive degeneration of the two successive motor neurons in the central nervous system, leading to a gradual weakening of all skeletal muscles of the body. Thus, all movements that use the extremities, the body trunk, but also facial muscles are affected. Patients initially note a clumsy hand or a weak leg; in a subset of patients the first symptoms are slurred speech and swallowing problems due to facial, oropharyngeal and neck muscle weakness (Box 12.1).

There are, to date, no therapeutic options that have the potential to cure ALS or halt its progression. The fact that a new and extremely expensive drug was approved after showing modest effectiveness in a subset of patients in a small and short-term pilot trial underscores the desperate need for a therapy to treat ALS [4]. In most cases, the disease inexorably leads to death within 3–5 years. Patients' lives can be sustained by mechanical ventilation, but at the same time the disease progresses and often leads to a locked-in state (LIS) that is characterized by a relatively intact mind that is locked in a completely paralyzed body, except for some eye and eyelid movements, which, however, may eventually also get lost ("complete LIS," CLIS).

Many health care professionals and researchers consider this to be a state that they would not want to live in and for which they would prefer forgoing life-sustaining measures [5]. What the patients in CLIS themselves think is unknown to us. A study among LIS patients who could still communicate reported that some maintained a high quality of life, while others felt miserable and preferred euthanasia [6]. We do not know what happens when LIS patients' last communication channel closes.

It is obvious that there is a very strong incentive to treat or mitigate the state of CLIS. Moreover, scientific and human curiosity drives our quest to know what it is like to live in such a state, without any possibility of expression and communication.

A technology that has the potential to be this desired window into CLIS is BCI. In the context of this book, it is not necessary to include a lengthy introduction into BCI [7]. It suffices to say that BCI is defined here as a technology that joins the human brain and a computer, thereby enabling a person to directly influence the environment via his or her own brain activity, without using the body's own motor system.

12.3 Ethical Implications of Medical Research Using Brain-Computer Interface

The ethical implications of medical research on BCIs is clearly exemplified by a recent case of alleged misconduct that happened in Germany. In 2017, the open-access peer-reviewed scientific journal "PLoS Biology" published an original article authored by Ujwal Chaudhary and colleagues, with the well-known neuropsychologist Niels Birbaumer as last author [8, 9]. The international author team (in addition to the German researchers there was also a scientist from the US National Institute of Neurological Disorders and Stroke, as well as scientists from China and Italy) was funded by a multitude of renowned funding agencies; among them the most reputable state funding agencies in Germany. Their paper, entitled "Brain-Computer Interface-Based Communication in the Completely Locked-in State," promised a major breakthrough in BCI research: for the first time in history, the authors contended to have successfully established communication with patients in CLIS.

Their publication disclosed the results of a case series of four patients that were reported to be in a CLIS. Using functional near-infrared spectroscopy (fNIRS), the researchers reported having decoded frontotemporal brain activity allowing them to distinguish between the mentalization of "yes" and "no" in response to orally presented questions of personal relevance (like the sentence "You were born in Berlin"). Three of the four patients were also asked so-called open questions about their attitudes towards their present situation and their life in general, yet the article was not very specific about these questions mentioning only that each patient was asked 40 open questions, such as whether they love their life or whether they feel sad.

The main summarizing statement by the authors was the following bold proposition:

Even after extended CLIS in ALS spanning months and years, reliable, meaningful communication using questions requiring a mental affirmative (yes) or negative, rejecting (no) answer is possible with fNIRS-BCI [8].

Regarding the open questions, they affirmed:

Patients F, G, and B answered open questions containing quality of life estimation repeatedly with a "yes" response, indicating a positive attitude towards the present situation and towards life in general (…) [8].

Strangely, the authors admitted that they did not have a plausible physiological explanation as to why the fNIRS responses were different between presumed "yes" and "no" responses.

Chaudhary et al. were aware that their experiments had an existential impact on the family members of the patients: "Family members of all four patients experienced substantial relief and continue to use the system" [8]. Interestingly, family members were always present during the experiments. The authors even mentioned that they never officially screened or recruited these patients, but it was the family members who approached the senior researcher and asked to participate in these experiments.

They were also aware of the ethical implications of their study:

Still, we have to remain cautious about our judgements to open questions' answers, particularly if it comes to quality of life and psychological changes of CLIS patients. In view of the gravity of the subject matter (i.e., establishing communication with nonverbal, completely paralyzed persons with preserved cognition), a call for replication of the current results by other investigators would be welcome [8].

This call for replication was repeated in the abstract of the article.

The media response to this article was overwhelming. Around the world, newspapers and news agencies reported about these experiments in an enthusiastic manner. "Decoding the thoughts of patients who can't even blink," was the CNN headline [10], "The Telegraph" used the headline "Locked-in patients tell doctors they are 'happy' after computer reading thoughts" [11], and the MIT Technology Review heralded the study as "Reached mind via a Mind-Reading Device, Deeply Paralyzed Patients Say They Want to Live" [12]. The scientific publication resonated with a long-held dream of mankind, the dream of mind-reading and of controlling the environment solely with thoughts [13, 14].

The researchers themselves appeared in the lay media presenting and explaining their research results, primarily the senior author Niels Birbaumer. In a long verbatim interview with the Germany weekly newspaper for intellectuals "Die ZEIT," he not only specified how the experiments were conducted and what the results mean, but also gave insights into his own personal motivations and attitudes [15] (Box 12.2).

Remarkably, Birbaumer mentioned that their results have enormous practical consequences for these families. One family obviously wanted to continue using the

Box 12.2: Extracts of a verbatim interview with Niels Birbaumer (Die ZEIT), translated by the author (RJJ) [15]

About his motivation to do this kind of research:

I am terrified about the situation of these people (…) No one makes any effort to do experiments with these unattractive patients (…) We do research and publish for 20 years. The result? Nothing (…) How many disappointments along the way! Without a few positive results now and then, I would have resigned long ago.

About the measurement of brain activity:

We pose the same question several times and the computer averages the yes/no responses (…) When the brain waves slow down, the computer does not count the answer (…) The computer sums up yes/no responses and calculates whether they correspond to the expected answers.

About the open questions:

Once we asked a patient whether his daughter would be allowed to marry her boy-friend. The answer, in 9 out of 10 cases, was no. Asked whether they are satisfied with their life, all responded yes.

About the continued BCI use after the end of the study:

One patient's family is using it regularly (…) We have already litigated together with the family of a patient (…) the court has decided that the € 50,000–70,000 device for locked-in patients has to be paid by the health insurance.

About the significance of their own results:

I now let other groups replicate the results. It would not be trustworthy if always the same researchers would do that.

BCI technology after the trial and had to litigate with the insurance company to receive a reimbursement of € 50,000–70,000, which is how much the device seems to cost, and the researchers appeared to be engaged in that litigation.

Birbaumer also reiterated the invitation from the journal article for colleagues to replicate the study. In fact, another researcher from the University of Tübingen, Germany, Martin Spüler, took this call seriously: while he did not replicate the whole study (an identical replication would be difficult for such a clinical case series), he took the published raw data and replicated the statistical analyses. His comment was published in the same journal as the original study several months later [16]. He claims to have used two different statistical models to analyze the data and did not find any significant difference in the fNIRS response between yes/no questions. He also exposed some methodological flaws of the original study. He basically asserted that the authors used a statistical calculation method that always leads to a significant answer, whatever the data may be.

In the same issue of the journal, the study authors refuted the critique by Spüler [17] and a researcher from England wrote a commentary that ended with a diplomatically worded criticism of the original publication:

> BCI research is interdisciplinary and is at the intersection of natural science, social science, engineering science, and medicine. Clear and simple communication is essential. Lack of detail can lead to confusion. Confirmation bias has an influence on the interpretation of results (…) To enhance clarity of communication, reports should (i) be written in simple language; (ii) methods should be clear, precise (…) and (iii) interpretation of results should be objective and realistic—in itself a hard task [18].

The whole issue went public through a newspaper report by scientific journalists in the renowned German daily newspaper "Süddeutsche Zeitung Magazin" [19], just a few days after the publication of Spüler's criticism. The report is a journalistic masterpiece under the heading of "Wunschdenken," in English "Wishful thinking." The three journalists claim to have talked to the persons involved and other expert researchers in the field, stating that experts who prefer to remain anonymous confirm the doubts about the quality of the study. Moreover, they have shed light on attempts of the study authors to silence their critical colleague. In fact, Spüler was fired by the University of Tübingen, the same university where the first and last study authors worked.

The press coverage, however, stirred international attention on this case [20] and prompted the University of Tübingen and the main funding agency to initiate investigations. A commission at the university confirmed the doubts about the scientific validity of this study [21, 22]. The main criticisms were (a) a highly selective choice of data for the publication (not publishing all data), (b) a lack of transparency concerning some data and the statistical methods, (c) missing data for results that were published in the article. The university asked the journal to withdraw the paper, which has eventually been done in December 2019, and informed the involved patients and insurance companies. Birbaumer himself criticized the university commission for lack of expertise. Later, the commission of the German national funding organization DFG confirmed this position and labeled the study as scientific fraud. It has banned the authors from applying and reviewing for the DFG for 5 years and has backed the withdrawal of the publications related to this study.

From a research ethics perspective, several issues are remarkable. First of all, this study underscores the fact that scientific research is never morally neutral. It is always embedded in a context of value-laden motivations of researchers, funders, and research subjects (or their families). These underlying moral assumptions and attitudes shape the choice of a research domain, the definition of study objectives, the selection of methods, and the interpretation of results. This case shows that it would be naïve to entertain the idea of a morally neutral, purely factual science. Rather, researchers should be aware of this moral dimension of their work and reflect on it in a self-critical and transparent way.

Second, this research example underlines that the psychology of the researcher may not always match the logic of research. While the latter should be the quest for falsification of a hypothesis (trying to disprove it), researchers are usually rather motivated by a strong conviction or even belief in their hypothesis, pushing them to try to prove the hypothesis at all costs. This attitude may lead to confirmation bias, influencing study design, methodology, and result interpretation so as to confirm the hypothesis. In fact, the highly competitive and commercialized research system favors this attitude: researchers have to deliver "positive" results to yield high-impact publications, funding, and attention in the scientific community and the public, necessary ingredients both for personal careers and the survival of research teams and centers.

A third implication of this case is the troubling observation that the rational criticism of study methods and results may be regarded by researchers as "whistle-blowing," whereas in fact it should be appreciated as the very essence and heart of scientific research. If critical thinking and expression become endangered, it will be the end of the kind of modern scientific research that the enlightenment has brought about and that proved to be so productive and successful. Small scientific communities working on rare diseases with very few patients might be particularly prone to this kind of danger, as all researchers in the field know each other and depend on each other for peer review of publications and funding applications.

A fourth implication concerns the technology more specifically. The controversy around this case is partly due to the complexity of the applied technology and the difficulty to retrace and understand the computer algorithms that were involved. In fact, as the researchers used a type of AI called deep learning algorithms, its characteristic is that it constantly changes its own method of data analysis in an automated way. In current deep learning devices, these automated adjustments may occur in a black box and may not be observable or retraceable one by one. In other words: using deep learning may confront us with research results whose scientific validity we may not be able to prove nor disprove. Did the BCI in the current case really detect true "yes" and "no" thoughts by the patients or was this result an artifact created by deep learning AI? The case may thus herald a major problem of AI that we will encounter more and more often with the increasing use of AI in the future.

Fifth, the case is troubling because it calls into question the role of research ethics committees. The article mentions that the internal review board of the University of Tübingen approved the experiments [8]. Did this board discuss the ethical implications of deep learning algorithms? How could it approve that such experimental technology was used to ask existential value-laden questions about the patients' will to live? In practice, these committees are often not equipped to identify and deliberate ethical questions, but rather limit themselves to applying checklists on methodology, informed consent procedures, and data protection.

Sixth, one of the ethical questions that such a committee should discuss is that of surrogate consent in studies with locked-in patients. Clearly, patients in a CLIS do not have the legal capacity to consent themselves to the study because functional communication is a prerequisite for this capacity [23]. According to the Declaration of Helsinki, which is the internationally accepted code of ethics on research with human subjects, experiments on these highly vulnerable patients are possible, but they have to respect additional safeguards [24]. Commonly, these studies should have the potential of a direct benefit to the study participant, e.g., an effective treatment of disease symptoms. Was this the case here? Was there a direct benefit from answering the yes/no questions? Or may these questions, at least in some patients, actually increase their awareness of their own suffering? Of course, one could argue that a well-functioning BCI communication device would be a huge advantage for patients in CLIS, but only once this technology is advanced enough to allow this kind of well-functioning, fine-grained communication. Thus, the study participants would only have what is called an indirect benefit: they, or other patients in a similar clinical situation, will benefit in the future, when the technology is ripe for application. Studies on subjects without decisional capacity that contain the potential for an indirect benefit, only however, may not have more than minimal harm and minimal burden according to the Declaration of Helsinki [24]. Whether undergoing fNIRS and the other investigations of the study satisfy the criterion of minimal burden is debatable.

The seventh and last point about the ethical implications of this study concerns the families of the involved patients. It is evident that families of such highly vulnerable, dependent patients who cannot act, decide, and communicate by themselves are highly involved in everyday care and decision-making. They often place very high hopes in technology, making them prone to becoming the victims of false promises and subsequent frustration [25]. When families contact researchers in order to enroll their patients into studies, as was the case here, this amounts to a self-selection of families with extremely high hopes, and researchers have to be very careful to avoid nourishing these exaggerated hopes.

In today's clinical reality, there is still not a single BCI application that has entered into routine health care. The most promising fields of medical BCI application in the near future are neurorehabilitation (restoring damaged motoneuronal pathways by pairing movement volition with external activation of the muscles or peripheral nerves), neurofeedback (training attention and mental focus using BCI tasks), and neuroprosthetics (controlling an arm or leg prostheses or even wheelchairs) [26–28]. As we have seen in the research project on patients in CLIS discussed above, most of these BCI applications will incorporate deep learning algorithms and AI that allow the BCI technology to learn from the person, adapt to her individuality and make movements and actions faster, more efficient and less effortful. Therefore, it is paramount to reflect as early as possible on the potential ethical and anthropological consequences and implications of AI in clinical care.

12.4   Clinical Use of Artificial Intelligence: Ethical and Anthropological Implications

BCI are one example of the use of AI in medicine. As in society in general, there are numerous potential applications for AI in health care. But in contrast to our everyday life, where we already use AI in smartphones, voice recognition devices, internet search engines, social media platforms, public transportation and many other domains, AI is still in its infancy in health care—even if the potential is vast. The most promising health care uses of AI can be seen in diagnostic and prognostic assessment, where large amounts of information have to be processed in order to increase the precision and validity of diagnostic or prognostic statements [29]. I will therefore first show some prime examples of diagnostic and prognostic AI uses and discuss their ethical implications afterwards.

Medical diagnosis relies more and more on imaging techniques and the visual recognition of pathological patterns. Typical examples are in dermatology, radiology, and endoscopy. A recent study has shown that a so-called neural network using deep learning paradigms could be trained so that its diagnostic performance in recognizing melanoma is actually superior to the performance of skilled dermatologists [30]. This kind of visual recognition may also function for diseases of the inner organs, and even the brain, that are associated with subtle changes in appearance: Fetal Alcohol-Spectrum Disorders could be reliably detected based on facial features by computer algorithms [31]. Moreover, AI can be a potent help in differential diagnosis, calculating the probabilities of various likely diagnoses based on a multitude of patient data [32].

In medical prognostication, physicians usually rely on validated scores, as well as on their intuitions that are ideally informed by long professional experience and many patient cases. Yet, this experience-driven knowledge is impossible with regard to rare diseases and hardly possible for junior physicians. Moreover, intuitions are prone to a host of psychological biases [33]. Thus, AI has the potential to significantly improve the accuracy of medical prognostication. As an example, an AI algorithm managed to predict survival rates based on microscopic pathology images in patients with non-small cell lung cancer [34]. One of the rarer studies where AI was used in medical treatment found that an AI-based chatbot that performed an automated form of an online cognitive behavioral therapy effectively reduced symptoms of anxiety and depression in young patients [35].

These few examples may already suffice to sketch the broad and diverse array of effective AI application in medicine, including neurology, coupled with BCI and other neurotechnology. Without exaggeration, one could expect a profound transformation of health care within a few decades. Diagnosis and prognosis will become much more precise and reliable. Disease entities will multiply as AI will help to differentiate between nuances of different diagnostic patterns as well as between different disease courses and responses to treatment. Moreover, refined diagnosis will also mean earlier diagnosis: AI may help to detect extremely subtle signs of diseases in a pre-symptomatic stage, increasing the prevalence of diseases in the population and contributing to a general trend of pathologizing and medicalizing our societies. The novel paradigm of medicine will be predictive medicine that can powerfully and precisely predict diseases, their course and symptoms, response to therapy, survival times, and many other characteristics—without necessarily having more to offer in terms of cure and effective treatment. The challenge for the patient is evident: How should he or she react to this novel kind of predictive knowledge? Will it restrain the degrees of individual liberty (at least subjectively perceived liberty) or even their autonomy? Will it stir a new kind of existential angst and lower quality of life or will we find ways to use it to increase quality of life? The main challenge will be to find positive, fruitful ways to deal with this new predictive knowledge in medicine.

The increasing use of AI in medicine will also challenge the patient–physician relationship, at least in three ways: first, it will reinforce the current trend of a growing interpersonal distance between the patient and the physician that already started with the invention of the stethoscope, deepened with radiology and is about to become even more distant by telemedicine. If the computer can diagnose, prognosticate and even treat better than a physician can, the latter will not need to go to the bedside; his place will only be at the computer, entering data, supervising the analysis, and interpreting the results. Ultimately, highly effective AI medicine could allow self-diagnosis and self-treatment from A to Z: the patient (or family members) could enter information and specimens, allow cameras and sensors to obtain data, receive an accurate diagnosis in the blink of an eye and get the appropriate medication a few minutes later by drone directly to his home. For a range of diseases and clinical situations, the involvement of a physician may not be necessary anymore. The ethical problem, however, is not only that the patient will become more and more isolated and lonely.

The dialogue between the patient and the physician that is so central to the acts of responsibility—defining a medical indication and obtaining informed consent—will be threatened. As the philosopher Emmanuel Levinas emphasized, the personal direct encounter of another human being (looking another person in the face) is what makes us responsible for our actions [36]. In this line of thinking, in AI medicine we cannot assume human responsibility, in a rich moral sense, in the same way as we can do when we have personal contact with the other person.

Second, AI medicine may lead to a desynchronization between the subjective time dimension of the ill person and the professional time dimension of medicine. The speed of AI algorithms is breathtaking and surpasses that of traditional medical diagnosis, prognosis, and treatment by far. Within a few seconds, a high-performing algorithm could find the correct diagnosis and define the treatment plan. It is questionable whether this saving of time really translates into valuable time for the persons involved, both the patient, the family, and the health care professionals [37]. Patients usually need many days, weeks, or even months to understand a diagnosis, to cope with a new life situation, and to make important treatment decisions. They also need other human beings with narrative capacity, who are able to understand their personal story, make sense of the development and course of a disease, and help to narrate an illness story that will be embedded into the life story of the patient [38].

Third, AI medicine has the potential to depersonalize health care in a profound way. Computer algorithms may have lots of advantages over physicians, but their inevitable shortcoming is that they always treat patients as instances of a collectivity, as statistical cases, and never as unique individual persons. This is inherent in their functioning. In fact, they do not understand the concept of a person because they themselves are not persons and lack human subjectivity. Yet, intersubjective relations are vital for human beings, especially for vulnerable persons with ill-nesses. In fact, it is these intersubjective encounters that mutually constitute our personhood. According to the German philosopher Axel Honneth, an individual cannot conceive himself as a genuine person without recognition by other individuals [39].

In summary, we have seen that AI may have profound consequences and challenges for health care. These challenges pertain to any kind of AI use in health care. Yet, if AI is applied to neurointerventions like BCI, the ethical problems of both areas converge and become even more troublesome. There is an inherent tendency in neurointerventions to use AI because they need to match the complexity of the human brain and this is best done by using deep learning algorithms that are based on artificial neuronal networks. I would even hypothesize that any new form of neurotechnological brain intervention that will be applied in humans will heavily rely on deep learning and AI. Yet, the inevitable combination of neurointerventional technology and AI for medicine potentiates the ethical problems. As I have shown, despite its remarkable performance, AI is in many respects alien to human needs and human characteristics: it cannot be take responsibility, it does not have emotions, it operates in a different temporal dimension, it follows a logic of depersonalization. When AI technology with these features is closely linked or even integrated into the human brain, this threatens what we call human nature, our human identity. The paradigm of the new, AI-based machine may merge with humans in such a way that what results could indeed be a novel kind of being, a cyborg that is not simply a combination, half human half computer, but a completely new entity in which human and technological parts may be indistinguishable. Thus, the fundamental distinction between human beings and technological artifacts might disappear in this new kind of entities.

Acknowledgment Work on this paper was funded by the Federal Ministry of Education and Research (BMBF) in Germany (INTERFACES, 01GP1622A).

References

1. Collaborators GBDN. Global, regional, and national burden of neurological disorders, 1990–2016: a systematic analysis for the Global Burden of Disease Study 2016. Lancet Neurol. 2019;18(5):459–80.
2. The Telegraph. Older people are more scared of dementia than cancer, poll finds. 2014. https://www.telegraph.co.uk/news/health/elder/11008905/Older-people-are-more-scared-of-dementia-than-cancer-poll-finds.html. Accessed 13 February 2021.
3. Stutzki R, Weber M, Reiter-Theil S, Simmen U, Borasio GD, Jox RJ. Attitudes towards hastened death in ALS: a prospective study of patients and family caregivers. Amyotroph Lateral Scler Frontotemporal Degener. 2014;15(1–2):68–76.
4. Writing G, Edaravone ALSSG. Safety and efficacy of edaravone in well defined patients with amyotrophic lateral sclerosis: a randomised, double-blind, placebo-controlled trial. Lancet Neurol. 2017;16(7):505–12.
5. Demertzi A, Jox RJ, Racine E, Laureys S. A European survey on attitudes towards pain and end-of-life issues in locked-in syndrome. Brain Inj. 2014;28(9):1209–15.

6. Bruno MA, Bernheim JL, Ledoux D, Pellas F, Demertzi A, Laureys S. A survey on self-assessed well-being in a cohort of chronic locked-in syndrome patients: happy majority, miserable minority. BMJ Open. 2011;1(1):e000039.
7. Wolpaw J, Winter Wolpaw E. Brain-computer interfaces: principles and practice. Oxford: Oxford University Press; 2012.
8. Chaudhary U, Xia B, Silvoni S, Cohen LG, Birbaumer N. Brain-computer interface-based communication in the completely locked-in state. PLoS Biol. 2017;15(1):e1002593 (Retracted article).
9. Chaudhary U, Xia B, Silvoni S, Cohen LG, Birbaumer N. Correction: brain-computer interface-based communication in the completely locked-in state. PLoS Biol. 2018;16(12):e3000089.
10. Howard J. Decoding the thoughts of patients who can't even blink. 2017. https://edition.cnn.com/2017/01/31/health/locked-in-als-brain-computer-study/index.html. Accessed 13 Febuary 2021.
11. Knapton S. Locked-in patients tell doctors they are 'happy' after computer reads thoughts. The Telegraph. 2017. https://www.telegraph.co.uk/science/2017/01/31/locked-in-patientstell-doctors-happy-computer-reads-thoughts/. Accessed 13 Febuary 2021.
12. Mullin E. Reached mind via a mind-reading device, deeply paralyzed patients say they want to live. MIT Technology Review. 2017.
13. Dehaene S. Reading in the brain: the science and evolution of a human invention. Viking: New York; 2009.
14. Seed D. Brainwashing: the fictions of mind control: a study of novels and films since world war II. Kent: Kent State University Press; 2004.
15. Landwehr T. Locked-in Patienten: "Ich bin entsetzt über die Situation dieser Leute". Die ZEIT. 2017. https://www.zeit.de/wissen/gesundheit/2017-01/locked-in-patienten-nielsbirbaumer-als-wachkoma-kopfhaube-kommunikation-schnittstelle. Accessed 13 February 2021.
16. Spuler M. Questioning the evidence for BCI-based communication in the complete locked-in state. PLoS Biol. 2019;17(4):e2004750.
17. Chaudhary U, Pathak S, Birbaumer N. Response to: "questioning the evidence for BCI-based communication in the complete locked-in state". PLoS Biol. 2019;17(4):e3000063.
18. Scherer R. Thought-based interaction: same data, same methods, different results? PLoS Biol. 2019;17(4):e3000190 (retracted article).
19. Bauer P, Illinger P, Krause T. Wunschdenken. Süddeutsche Zeitung Magazin. 2019. https://sz-magazin.sueddeutsche.de/gesundheit/wissenschaft-als-nervenkrankheit-professorniels-birbaumer-forschung-gehirn-medizin-87138?reduced=true. Accessed 13 Februry 2021.
20. Vogel G. Research on communication with completely paralyzed patients prompts misconduct investigation: AAAS. 2019. https://www.sciencemag.org/news/2019/04/research-communication-completely-paralyzed-patients-prompts-misconduct-investigation. Accessed 13 February 2021.
21. Neuroskeptic. The fall of Niels Birbaumer. 2019. https://www.discovermagazine.com/the-sciences/the-fall-of-niels-birbaumer. Accessed 13 February 2021.
22. Hütten F, Krause T. Kommission sieht wissenschaftliches Fehlverhalten im Fall Birbaumer Süddeutsche Zeitung. 2019. https://www.sueddeutsche.de/gesundheit/deutschlandtuebingen-als-birbaumer-gehirn-gedanken-clis-1.4477468. Accessed 13 February 2021.
23. Appelbaum PS. Clinical practice. Assessment of patients' competence to consent to treatment. N Engl J Med. 2007;357(18):1834–40.
24. World Medical Association. Declaration of Helsinki, revision 2013. 2013. https://www.wma.net/policies-post/wma-declaration-of-helsinki-ethical-principles-for-medical-research-involving-human-subjects/. Accessed 13 February 2021.
25. Jox RJ, Bernat JL, Laureys S, Racine E. Disorders of consciousness: responding to requests for novel diagnostic and therapeutic interventions. Lancet Neurol. 2012;11(8):732–8.
26. Cervera MA, Soekadar SR, Ushiba J, Millan JDR, Liu M, Birbaumer N, et al. Brain-computer interfaces for post-stroke motor rehabilitation: a meta-analysis. Ann Clin Transl Neurol. 2018;5(5):651–63.

27. Jeunet C, Lotte F, Batail JM, Philip P, Micoulaud Franchi JA. Using recent BCI litera-
    ture to deepen our understanding of clinical neurofeedback: a short review. Neuroscience.
    2018;378:225–33.
28. Shih JJ, Krusienski DJ, Wolpaw JR. Brain-computer interfaces in medicine. Mayo Clin Proc.
    2012;87(3):268–79.
29. Hosny A, Parmar C, Quackenbush J, Schwartz LH, Aerts H. Artificial intelligence in radiol-
    ogy. Nat Rev Cancer. 2018;18(8):500–10.
30. Haenssle HA, Fink C, Schneiderbauer R, Toberer F, Buhl T, Blum A, et al. Man against machine:
    diagnostic performance of a deep learning convolutional neural network for dermoscopic mel-
    anoma recognition in comparison to 58 dermatologists. Ann Oncol. 2018;29(8):1836–42.
31. Valentine M, Bihm DCJ, Wolf L, Hoyme HE, May PA, Buckley D, et al. Computer-
    aided recognition of facial attributes for fetal alcohol spectrum disorders. Pediatrics.
    2017;140(6):e20162028.
32. Cahan A, Cimino JJA. Learning health care system using computer-aided diagnosis. J Med
    Internet Res. 2017;19(3):e54.
33. Norman GR, Monteiro SD, Sherbino J, Ilgen JS, Schmidt HG, Mamede S. The causes of errors
    in clinical reasoning: cognitive biases, knowledge deficits, and dual process thinking. Acad
    Med. 2017;92(1):23–30.
34. Yu KH, Zhang C, Berry GJ, Altman RB, Re C, Rubin DL, et al. Predicting non-small cell lung
    cancer prognosis by fully automated microscopic pathology image features. Nat Commun.
    2016;7:12474.
35. Fulmer R, Joerin A, Gentile B, Lakerink L, Rauws M. Using psychological artificial intel-
    ligence (Tess) to relieve symptoms of depression and anxiety: randomized controlled trial.
    JMIR Ment Health. 2018;5(4):e64.
36. Levinas E. Ethics and infinity. Pittsburgh: Duquesne University Press; 1995.
37. Rosa H. Social acceleration: a new theory of modernity. New York: Columbia University
    Press; 2015.
38. Frank AW. Narrative ethics as dialogical story-telling. Hast Cent Rep. 2014;44(1 Suppl):
    S16–20.
39. Honneth A. Integrity and disrespect: principles of a conception of morality based on the theory
    of recognition. Political Theory. 1992;20(2):187–201.

Practical, Conceptual and Ethical Dimensions of a Neuro-controlled Speech Neuroprosthesis

13

Stephen Rainey

Contents

Abstract

A speech neuroprosthesis is technology that is designed to pick out linguistically relevant neural signals. These can be recorded from the parts of the brain that are associated with speech comprehension and production. Relevant signals can be derived from the surface of the brain or deeper within it. Different areas and different recording techniques (e.g. surface vs. intracortical probes) provide different ways of accessing linguistically relevant signals. The principle of each approach, however, is centred on decoding the neural activity so that the overt speech sounds they represent can be synthesised and realised artificially. This raises the possibility of restoring speech for those who may have lost their ability to communicate through, for example, disease or injury. This raises potential ethical questions, especially given the role of artificial intelligence in decoding approaches. These issues address to what degree synthesised speech, via the prosthesis, represents the user's speech intentions, or their perspective more generally. Important questions about user control of speech neuroprostheses are addressed here.

S. Rainey (✉)
Oxford Uehiro Centre for Practical Ethics, University of Oxford, Oxford, UK
e-mail: stephen.rainey@philosophy.ox.ac.uk

© Springer Nature Switzerland AG 2021
O. Friedrich et al. (eds.), *Clinical Neurotechnology meets Artificial Intelligence*,
Advances in Neuroethics, https://doi.org/10.1007/978-3-030-64590-8_13

13.1   Introduction

Neurotechnology, in general, is technology operated entirely by way of recording, processing, decoding and instrumentalising neural signals. Brain-computer interfaces (BCIs) intended to reproduce speech are a growing area of interest within this context. Neuroprostheses for the production of speech in cases of profound paralysis, which can result from amyotrophic lateral sclerosis (ALS) for example, are being developed. The Horizon 2020-funded BrainCom project is a multidisciplinary European project that is working on developing neuroprosthetics for speech that build on this neurotechnological principle of instrumentalising neural signals [1].

A speech neuroprosthesis is technology that is designed to pick out linguistically relevant neural signals. These can be recorded from the parts of the brain that are associated with speech comprehension and production. This might include areas associated with the movement of lips, tongue, throat and other articulators. Relevant signals can be derived from the surface of the brain or deeper within it. Different areas and different recording techniques (e.g. surface vs. intracortical) provide different ways of accessing linguistically relevant signals. The principle of each approach, however, is centred on decoding the neural activity so that the overt speech sounds they represent can be synthesised and realised artificially. This raises the possibility of restoring speech for those who may have lost their ability to communicate through, for example, disease or injury.

One specific approach to this is based in the articulatory motor neural correlates of overt speech, as mentioned. When users of a speech neuroprosthesis vividly imagine that they are saying words—when they say the words clearly in their head—the articulator-relevant areas of the brain release signals that are similar to those that occur when saying those words out loud [2–4]. They target specific recording sites that are known to be active during speech production [5–7]. One set of these sites comprises the motor areas associated with speech articulators: the jaw, lips, tongue and so on. Recording the electrical activity of these areas while the user concentrates on imagining speech allows the neural activity closely correlated to that which is produced when speaking out loud to be recorded. Through this targeted use of neural recording, and sophisticated deployment of signal processing, speech production can be predicted. This is how the articulatory area approach opens the possibility of externalising unvoiced speech for language-compromised users.

Interesting questions can arise with specific reference to the nature, degree and role of processing neural signals so that speech is the overall output. In particular, artificial intelligence (AI) has an interesting role in making decoding strategies efficient and realistic, but might raise questions in terms of overshadowing the intentions of the putative speaker. It will be worth exploring some themes that arise in neurotechnology in general before moving on to those concerning speech devices.

13.2   Neuro-controlled Technology

An electroencephalogram (EEG) electrode placed against the scalp can detect electrical activity within the brain. The activity that is most clearly detectable is that closest to the electrode. Activity occurring deeper within the brain becomes harder to distinguish among the other general signals. There is a limitation in EEG resolution so that specific electrical activity cannot be targeted easily, since it is all happening at once. Brain signals are easily recorded, but to discern specific signals from the general buzz requires clever processing work.

Nevertheless, EEG provides sufficient clarity to be able to drive a variety of neurotechnologies. More invasive techniques can be deployed where greater resolution is required. These include intracerebral EEG (iEEG) or microelectrocorticography (μECoG). This latter technique provides greater resolution in that it is located on the surface of the brain rather than the skull, thereby recording from specifically targeted sites. If EEG is like holding a microphone to a closed door and discerning the muffled sounds within, these latter techniques are like having one in the room.

Signals thus recorded can be processed by computer in order to drive neuro-controlled devices or applications of various type, such as prosthetic limbs or software programs. Given the nature of these brain-computer interfaces (BCIs), it might appear reasonable to argue that they present different conditions of moral and ethical evaluation as compared to conventional action cases. The 'acts' that constitute BCI-mediated action are mental acts, which do not require physical bodily action. Factors such as freedom of thought ought to mean that we treat these actions differently from conventional actions. The law appears to back up this view by limiting responsibility for actions based on notions to a 'willed bodily movement' (the 'act requirement', as discussed in [8]).

BCI-mediated action offers unique ways in which intentional human action may be realised in the material world without an apparent act. This is unique in that it is non-hormonal and non-muscular [9], which have hitherto been the only means of humans effecting intentional action in the world. Jonathan Wolpaw describes the signals users are required to learn to operate: '...not normal or natural brain output channels' [10]. Together, these observations suggest the general peculiarity of BCI-action from the perspective of external action and 'internal', neural control.

In the standard case of neural correlates for actions, questions of neural control do not seem to arise. Wondering about a person's intentions when they move their arm typically does not include wondering about their neurons. In the BCI case, neural control is very much at stake. In principle, the application of scientific methods on human action might even undermine taking responsibility for these actions, or more specifically, having control over these actions. If we begin reducing actions to a neural basis and not an intentional one, we might lose sight of 'actors' being involved in their own acts.

A majority of the control systems of the human body that relate to balance, digestion, heart rate, perception and so on, are automatic in the sense that they are constituted by neural activity that is not in direct control of the person. This essentially means that electrical activity is constantly occurring inside the skull and serves to regulate a large amount of bodily activity. Detectable neural activity is electrical activity, which occurs when neurons fire. To repeat the analogy from above, EEG can record this activity much like a microphone held up to a door can record sound from the room next door.

Despite the fact that so many neural processes happen automatically, the kinds of brain signals that can be detected, recorded and decoded can reveal a lot about the state of the body. From these revelations, inferences can be drawn about things like mood, memory, motor intentions or even seemingly complicated notions that have to do with taste and intention [11–18]. These signals can reveal dimensions of cognitive activity or evidence of complex thoughts. In a more practical dimension, they can also drive various devices and software programs. These kinds of applications can permit a greater scope of activity to device users who have mobility or other impairments. For example, neural recordings can be decoded to control a wheelchair. They may also be used as input for a speech prosthesis.

It might seem that actions performed by using the brain to drive a device of some kind are special cases of action. It is true that they are not conventional cases. Should they be considered radically different to conventional actions? After all, physical movement is not required in order to use a neural-controlled device, whereas bodily movement is usually a marker of action. Moreover, an action is very commonly ascribable to an agent because they are visibly performing it. Here, with neuro-controlled action, this ascription condition is not clear. Centrally, where action is brought about without bodily movement, but through the realisation of neural patterns of activity, should it be thought of as mental action? Even though there are physical world effects, it might seem tempting to characterise the kinds of responsibility for action brought about through a BCI-controlled device as relating to freedom of thought, rather than conventional action. If this line of thought were followed, it might seem difficult to account for real-world consequences of neural-controlled actions. This would be unfortunate.

On a belief, desire and intention model of action [19, 20], reasons emerge as a way to realise one state of affairs from another. If I believe the water in front of me is potable, and I am parched, I reach for it to slake my thirst. Reaching for and taking the water is caused 'in the right way' as it is done for the reason of thirst-quenching. This kind of action is conventional, and fairly transparent to the actor. Introspection of the neural activity driving devices is not possible, but access to one's own reasons is unimpaired. Whether the action is completed by way of a bodily arm, a prosthetic limb or some more elaborate system, the action on this account is the same. Reasons are what matter as they are what, from an agent perspective, mark activity as *action*.

Neurotechnologies aimed at decoding speech from neural signals are an example where this is of particular importance. Owing to the complexity of language and speech behaviour, reasons for saying one thing as opposed to another are vital. The content of speech, the very meaning of words and sentences, depends on the reasons that can be offered for having said those very words and sentences.

Were an argument to be judged on the basis only of its conclusion, there would likely be a lot of interesting thought lost. Evaluating arguments means looking at the reasons offered that lead to conclusions. Bad arguments should be rejected even where conclusions embody good propositions. Likewise, good arguments for bad propositions ought to be challenged. This applies not only to formal logical arguments, or language in the abstract, but to the phenomenon of language in general. In order to see how reasons can be preserved in the case of a neural speech prosthesis, the specifics of neural-signal driven speech devices require some further description.

13.3   Neural-Signal Driven Speech Devices

General neural-recording technologies, such as functional magnetic resonance imaging (fMRI) or EEG, are relatively effective at indicating the sorts of neural signals relevant to speech [2, 21–23]. In terms of fMRI, these indications are based on detectable blood oxygen levels. Because neural activity consumes oxygen, depleted oxygen levels in the blood indicate neural activity. However, this operates too slowly and generally at a resolution that is too low to be of use in a realistic speech situation, such as having a conversation. EEG, while capable of operating more quickly, faces a bottleneck in terms of signal differentiation. Because the signals reaching the electrode are a sum of a variety of neural sources at varying distances from the skull, a challenge emerges in disentangling one signal from another. Progress in μECoG recording has provided the means to surpass these limitations [24–28].

Via probes placed directly onto the cerebral cortex, the surface of the brain, high-resolution electrical activity can be read from the brain very quickly. These probes can be placed in important regions of the cortex that are related to speech. BrainCom is developing technology based on this kind of approach by decoding articulatory-related activity from premotor, motor and Broca cortex areas. These are brain regions that are associated with planning movement, undertaking movement and linguistic expression. From the information gleaned using sophisticated electrode arrays, speech can be decoded with very high accuracy. This is based on motor-area activity, processed so as to indicate the phonemes that such activity would create if it were to move the articulator muscles. Neural signals indicate the likely movements of the articulators and thus provide a basis to infer mouth position, lip shape and so on, and thus infer sounds.

Articulatory motor neural activity associated with speech can remain intact even without the articulators having motor function. In externalising the neural activity by artificial means, the paralysed body is effectively bypassed and speech activity is made possible. Appropriately processed and decoded, this information can be used in an artificial speech system. Ultimately, the speech that the prosthesis user would have spoken can be relayed synthetically based on the neural activity recorded [24, 29].

13.4   Risk Factors with Intracortical Probes

Among the benefits of fMRI or EEG are their non-invasiveness. Although undergoing an fMRI scan may not be entirely comfortable in that one is confined in a narrow plastic tube, it nonetheless does not require physical intervention. Another similar imaging technique, positron emission tomography (PET), does require the injection of radioactive isotopes in order to create an image. The images gained from fMRI are generated through the disturbance of protons within hydrogen via radio waves, and so do not impact bodily integrity. EEG, too, involves only the wearing of an electrode-studded cap, meaning that the body remains intact.

High-density electrode arrays and intracortical probes present the spatio-temporal resolution benefits by way of their proximity to the site of neural activity. This involves that they are implanted beneath the skull on the surface of the brain or deeper within the brain. This clearly involves highly invasive surgery with associated risks [30]. Some risks occur from the onset given the non-trivial nature of opening the skull to place the electrodes. Post implantation, the chance of infection increases in that the probes within the skull require outputs that cross the skin barrier. These infections are particularly serious seeing as they are so close to the brain. This raises very practical concerns about how to create, treat, maintain and monitor sites where electronics, the scalp, skull and brain interface.

These risks become particularly ethically important when decisions must be made about who ought to be considered for such procedures. Given the very serious nature of the surgery, at least a proportional payoff must be present in order to justify the risk. No one would consider such drastic intervention for a minor speech condition. But in cases of locked-in syndrome (LIS) for instance, where a patient may be physically paralysed but cognitively intact, the recovery of a robust means of communication might be seen as a sufficient benefit. Such a recovery is linked not only to improved communication, but also improvement in life expectancy [31]. Communication is possible for many, even the most seriously paralysed, but it can be cognitively taxing, slow and prone to error. The use of an assistant with a letter board can be very effective, but essentially means that independent communication is not available [32]. Eye-blink or EEG-based software control, as opposed to word-processing technology for instance, presents more independence, but is not appropriate for every patient and can be difficult to master [10, 33].

Only in very serious cases of communication impairment with the biggest likelihood of success and benefit would surgery be thought of as appropriate. The recovery of a communicative ability in a patient with LIS would be a good and clear representative example of when this might be appropriate. Apart from these important ethical considerations relating to the use of surgery and invasive devices, other ethical issues can arise in terms of the speech system itself—how it operates and its contexts of use. These will be explored next, beginning with the role of neural-signal processing.

13.5   Processing Brain Signals for Speech

Neural activity occurs within milliseconds. Meanwhile, normal conversational speech activity is also very fast. The technical challenge of recording, processing and decoding neural signals associated with speech in order to produce real-time synthetic speech is quite substantial. This is all the more acute when a normal conversational rate is something that is trying to be reproduced artificially.

Appropriate processing and decoding might by necessity have to include a predictive element in order to keep up the pace. This might be '…a statistical language model (giving the prior probability of observing a given sequence of words in a given language)' [2]. This kind of model aids in word prediction and aims '[t]o capture important syntactic and semantic information of language… by calculating probabilities of single words and probabilities for predicting words' [25]. The inclusion of this kind of element in the overall system ought to make it work faster and more reliably.

The technical benefits of a language-recognising prediction element within the architecture of a 'brain-to-speech' processor could have an obscure but potentially important side effect associated with it. Word distribution in natural languages approximately follows a Zipfian distribution. This means that while some words appear very frequently, most are very rare [34]. The processing demands of dealing with this type of distribution are very high, especially if some form of artificial intelligence is not included.

The mediation of neural activity and the use of prostheses means that the effort must be made to 'learn a new language' [35] and to 'use the brain' [36] to trigger and then control the device. In this sense, it appears that the device is an instrument that is being put to use [10]. However, given that this triggering and control is modified by some form of AI language-predicting software, it would appear as though the device has an element of activity that is outside the scope of user control.

Neural activity processing software may operate according to both generic rules, as discovered in a statistical model of language, and certain rules developed from interfacing with a specific user. A kind of delegation-of-action element seems to be at work in such an arrangement [37, 38]. Yet, this seems to raise the potential of tensions between the faithful reproduction of articulatory motor signals as recorded from the brain and grammatical phrase prediction. To what extent is the prosthesis a tool that is used by the would-be speaker, and to what extent is it a semi-autonomous predictive machine?

The answer to this question might depend on the sophistication of the software that is being used. That, described in Herff et al. [25], appears to be based on the statistical likelihood of two words co-occurring in specific texts. This is also complemented by the conditional probability of a word occurring through particular neural activity that is associated with linguistic (phonetic, syllabic or other) features. Using artificial neural nets in language neuroprostheses in general as a control feature [24, 39], a more sophisticated language model that is based in an artificial neural net might change things. Such a net could offer the improvements in prediction given by the word co-occurrence approach, yet with more agility in terms of actual spoken language [40, 41].

All artificial neural nets can learn from example data and can make generalisations, or they can learn from given examples about how to treat novel cases [42, 43]. In so doing, however, they risk becoming like 'black boxes'. This may be the case in the sense that the specific rules they employ at a given time can appear opaque to the outside observer. If a neuroprosthetic speech device were to use a sophisticated model of language that could respond to novel data by generalising according to rules learned from training data, this could amount to the sort of semi-autonomous predictor problem indicated above. It would seem possible that a tension could emerge between actual linguistic intentions as realised in neural activity, and the very sophisticated predictions of the model.

The nub of the issue would be how to discern when a model appropriately augments a neural recording to reflect better user speech, and when it departs from user intent to an unacceptable degree. For example, it would not be appropriate for the model to use a small amount of neural signals as input only to then overshadow the decoding process with its predictive function. This would divert too greatly between the synthetic speech output and the reasons the agent holds for speaking at all, which goes back to the Davidsonian account of action mentioned above.

If a sophisticated language-prediction model became central to a speech neuroprosthesis system, issues over the responsibility for the realised synthetic speech would become salient. Insofar as the balance between a speaker's reasons and intent reflected in neural signals, and prediction software are thrown into question it is important not to overlook the hybrid nature of specific speech output. An unbalanced relation between prosthetic device and user could not only put speaker intent at stake, but also agency more generally. We may end up without a clear means of expecting the user's reasons to be what rationalised the action, and not the prediction model.

Machine-learning researchers are presently engaged in finding ways to make systems account for their own actions in some way, as discussed in Doshi-Velez and Kim [43]. In some cases, given sufficient developments in machine-learning technology, there could be at least two accounts in terms of reasons for *this* speech output as opposed to *that* speech output. Overall, it might be unclear as to whether speech ought to count as the speaker's or the system's action, or a hybrid or something else.

Exacerbating this issue, a phenomenon known as 'verbal overshadowing' may also come into play. This phenomenon occurs when sincere reports made by an individual are subsumed by 'objective' descriptions given to them from an external source [44]. Eyewitness reports, for example, can be skewed away from accuracy when a witness receives specific details about their testimony from a third party. The objectivity of synthesised speech output could overshadow user intent in the same manner.

A neuroprosthesis user could, upon hearing synthesised speech that is close to their intent, accept it as their own. Incrementally, the predictions of the system could gain prominence, from the user point of view, by hearing synthesised speech as a reason to accept that very speech as one's own. *Who said what* would then arise in a very subtle and tricky way. This might be seen as a mirror of various concerns raised in terms of deep brain stimulation and about freedom and responsibility in action following neural interventions [45–47].

Even on a neurophysiological basis this can manifest as an issue. The superior temporal gyrus (SPG) is sensitive to the phonetic features of heard speech as well as being active in speech production [21, 48]. This could serve to muddy the waters as to where speaker intent and speaker perception coincide. The speech system, or the users themselves, could become confused over what was being intended as speech and what was heard as playback, given an overlap in the SPG [49], especially in cases where medication or cognitive issues may play a role.

A neuroprosthetic speech device must likely rely upon AI in its decoding of neural signals in order to be fast enough to reproduce recognisably conversational language. The extent to which that AI decoding utilises a language model is of importance in order to be able to separate the language intentions of the neuroprosthesis user from the decoding features of the system overall. The output from the neuroprosthesis can be described as 'hybrid', in some sense at least, by drawing upon speaker intentions, reasons and sophisticated processing. Maintaining a balance so that language outputs can be confidently ascribed to the user in every case is not a trivial matter. This can be framed in terms of a question of control, which will be addressed next.

13.6 Control over Synthetic Speech

Control over a neural signals-based device can, to the outside observer at least, appear quite obscure. It is a relatively new phenomenon to be able to control devices without any movement of the body. This represents a novel means of human action in a non-bodily sense. This raises some questions concerning control, especially with regard to communication.

How can we be sure that the device user appropriately controls a device? In neurotechnology generally, recorded signals that control a device might be checked against what a user reports as their intention. For instance, if someone was operating a neural-controlled drone that flew to the left, we could ask if that was what they were trying to do. The user may or may not concur. Where communication is at stake, though, there is an extra level of complexity precisely because we cannot ask about user intentions in operating the device except via the very device in question.

It should be noted that training with a neuro-controlled device is essential. It has been described as being like learning a new language, where the control of a prosthetic limb is concerned [35]. Moreover, in that a device itself adapts to its user, a kind of hybrid co-adaption occurs in which the user becomes acquainted with the device, and the device changes to suit the user [50]. This suggests some of the potential areas for questioning the nature and role of the device being in control where BCIs in general are concerned.

How could we best imagine a system that permits a high degree of confidence in a speech prosthesis as appropriately representing the speech of a user? Especially when the speech system is coupled to the articulatory motor areas of the brain, it seems unlikely that synthetic speech could stray far from intended speech. The articulatory motor cortex correlates of a plosive *t* sound, for instance, are unlikely to be misread or misrepresented as a fricative *f*. The different physical positions of the articulators correlate with different neural activity and so, at this level, misidentification is minimal.

At another level, interesting issues may arise owing to the nature of covert speech as a control element of the device. The vivid imagination of speech 'in the head' produces neural activity in the articulatory motor cortex that is similar to the activity generated in speaking out loud. This is what is exploited in a neuroprosthesis for speech, of the type discussed here. It is not impossible that vividly imagined speech could occur even though it was not intended to be spoken out loud. After all, in one sense at least, the kind of speech prosthesis up for discussion here is designed to predict overt speech based in covert speech signals. The reaction of the device to signals is not constrained by intention, but by signal features. A random thought, or a strongly felt reaction, might manifest as covert speech if it has the right signal form. This seems to raise the possibility that user control over a speech prosthesis might be compromised.

Not only might something be vividly imagined that was not intended for overt speech, it might even misrepresent the perspective of the speaker. Thoughts that we do not believe in can easily occur to us merely from the free play of imagination, or from memory, or mind-wandering. This seems to be an ethical problem in at least two ways. Firstly, in terms of sufficient control by the user over the speech device. Without sufficient control, a user may be unable to appropriately relate their speech intentions. Secondly, through the use of the device the user could be misrepresented by having synthetic speech produced that did not cohere with their values and perspective.

One main way to deal with these issues immediately presents itself. A means for the user to have a form of veto over their speech output might serve well to mitigate issues arising from these concerns. In getting to listen to the output before a wider audience, the user could veto the final production of synthesised speech if it does not coincide with what they intended, or was not intended at all. This would have the benefit of dealing with the issues that have been raised, and doing so via technical means. The downside might be that the extra step may slow down the pace of conversation.

Another more social approach would rely upon the audience to seek verification of the speaker's intentions and speech contents by pragmatic means. For instance, if the user of a speech prosthesis were to make a claim, the audience could ask questions about that claim in ways that minimised possible ambiguity and provided scope for retraction where views not held were expressed. At any rate, this is not an unfamiliar part of conventional communication. From time to time we seem to find out what we mean through the speech we produce. Or, put differently, we don't necessarily have a clear plan about what exactly we will say next when in conversation, but we find that out as we go along. An extension of this kind of practice in cases of conversation via neuro-controlled speech devices might not be too difficult to adapt to.

These last issues still lay ahead in the future. At present, the technology required to create real-time neuroprostheses for general conversation is in its infancy. Yet, since it is being developed now, it is the right time to provide reasonably constrained speculation about the potential for emerging ethical issues. By providing analysis early, good solutions can be integrated from the start and the systems that ultimately emerge will be better for it.

Acknowledgment   The author gratefully acknowledges funding from the BrainCom Project, Horizon 2020 Framework Programme (732032).

References

1. BrainCom: collaborative research project, neurorehabilitation. BrainCom. http://www.braincom-project.eu/. Accessed 19 Feb 2019.
2. Bocquelet F, Hueber T, Girin L, Chabardès S, Yvert B. Key considerations in designing a speech brain-computer interface. J Physiol Paris. 2016;110(4):392–401. http://linkinghub.elsevier.com/retrieve/pii/S0928425717300426. Accessed 1 Nov 2017.
3. Geva S, Jones PS, Crinion JT, Price CJ, Baron J-C, Warburton EA. The neural correlates of inner speech defined by voxel-based lesion-symptom mapping. Brain. 2011;134(10):3071–82. https://academic.oup.com/brain/article-lookup/doi/10.1093/brain/awr232.   Accessed   12 Jun 2017.
4. Mugler EM, Patton JL, Flint RD, Wright ZA, Schuele SU, Rosenow J, et al. Direct classification of all American English phonemes using signals from functional speech motor cortex. J Neural Eng. 2014;11(3):035015. http://stacks.iop.org/1741-2552/11/i=3/a=035015. Accessed 12 Jun 2017.
5. Chakrabarti S, Sandberg HM, Brumberg JS, Krusienski DJ. Progress in speech decoding from the electrocorticogram. Biomed Eng Lett. 2015;5(1):10–21. http://link.springer.com/10.1007/s13534-015-0175-1. Accessed 12 Jun 2017.
6. Guenther FH, Brumberg JS, Wright EJ, Nieto-Castanon A, Tourville JA, Panko M, et al. A wireless brain-machine interface for real-time speech synthesis. PLoS One. 2009;4(12):e8218. http://journals.plos.org/plosone/article?id=10.1371/journal.pone.0008218.   Accessed   22 Aug 2017.
7. Jarosiewicz B, Sarma AA, Bacher D, Masse NY, Simeral JD, Sorice B, et al. Virtual typing by people with tetraplegia using a self-calibrating intracortical brain-computer interface. Sci Transl Med. 2015;7(313):313ra179.
8. Munoz-Conde F, Chiesa LE. The act requirement as a basic concept of criminal law. Cardozo L Rev. 2006;28:2461.
9. Steinert S, Bublitz C, Jox R, Friedrich O. Doing things with thoughts: brain-computer interfaces and disembodied agency. Philos Technol. 2019;32:457–82. http://link.springer.com/10.1007/s13347-018-0308-4. Accessed 6 Aug 2018.
10. Wolpaw JR, Birbaumer N, McFarland DJ, Pfurtscheller G, Vaughan TM. Brain–computer interfaces for communication and control. Clin Neurophysiol. 2002;113(6):767 91. http://www.sciencedirect.com/science/article/pii/S1388245702000573.
11. Vyas V, Saxena S, Bhargava D. Mind reading by face recognition using security enhancement model. In: Proceedings of fourth international conference on soft computing for problem solving. New York: Springer; 2015. p. 173–80.
12. Wexler A, Thibault R. Mind-reading or misleading? Assessing direct-to-consumer electroencephalography (EEG) *devices marketed for wellness and their ethical and regulatory impli-*

*cat*ions. J Cogn Enhancement. 2019;3:131–7. https://doi.org/10.1007/s41465-018-0091-2. Accessed 10 Oct 2018.

13. Sani OG, Yang Y, Lee MB, Dawes HE, Chang EF, Shanechi MM. Mood variations decoded from multi-site intracranial human brain activity. Nat Biotechnol. 2018;36(10):954–61. https://www.nature.com/articles/nbt.4200. Accessed 16 Oct 2018.

14. Vincent NA. Neuroimaging and responsibility assessments. Neuroethics. 2011;4(1):35–49. https://doi.org/10.1007/s12152-008-9030-8. Accessed 2 Nov 2018.

15. Meegan DV. Neuroimaging techniques for memory detection: scientific, ethical, and legal issues. Am J Bioethics. 2008;8(1):9–20. http://www.tandfonline.com/doi/abs/10.1080/15265160701842007. Accessed 16 Oct 2018.

16. Calder AJ, Lawrence AD, Young AW. Neuropsychology of fear and loathing. Nat Rev Neurosci. 2001;2(5):352–63. https://www.nature.com/articles/35072584. Accessed 1 Feb 2019.

17. Haynes J-D, Sakai K, Rees G, Gilbert S, Frith C, Passingham RE. Reading hidden intentions in the human brain. Curr Biol. 2007;17(4):323–8.

18. Steinert S, Friedrich O. Wired emotions: ethical issues of affective brain–computer interfaces. Sci Eng Ethics. 2020;26:351–67. https://doi.org/10.1007/s11948-019-00087-2. Accessed 15 Mar 2019.

19. Davidson D. Subjective, intersubjective, objective: philosophical essays volume 3. Oxford: Oxford University Press; 2001. http://ebookcentral.proquest.com/lib/oxford/detail.action?docID=3052646. Accessed 16 Aug 2018.

20. Davidson D. Essays on actions and events: philosophical essays. Oxford: Oxford University Press; 2001. http://ebookcentral.proquest.com/lib/oxford/detail.action?docID=3052652. Accessed 16 Aug 2018.

21. Buchsbaum BR, Hickok G, Humphries C. Role of left posterior superior temporal gyrus in phonological processing for speech perception and production. Cogn Sci. 2001;25:663–78.

22. Price CJ. A review and synthesis of the first 20 years of PET and fMRI studies of heard speech, spoken language and reading. NeuroImage. 2012;62(2):816–47. http://www.sciencedirect.com/science/article/pii/S1053811912004703.

23. Zheng ZZ, Munhall KG, Johnsrude IS. Functional overlap between regions involved in speech perception and in monitoring one's own voice during speech production. J Cogn Neurosci. 2010;22(8):1770–81. http://www.ncbi.nlm.nih.gov/pmc/articles/PMC2862116/.

24. Bocquelet F, Hueber T, Girin L, Savariaux C, Yvert B. Real-time control of an articulatory-based speech synthesizer for brain computer interfaces. PLoS Comput Biol. 2016;12(11):e1005119. http://journals.plos.org/ploscompbiol/article?id=10.1371/journal.pcbi.1005119. Accessed 17 Apr 2018.

25. Herff C, Heger D, de Pesters A, Telaar D, Brunner P, Schalk G, et al. Brain-to-text: decoding spoken phrases from phone representations in the brain. Front Neurosci. 2015;9:217. https://www.frontiersin.org/articles/10.3389/fnins.2015.00217/full. Accessed 8 May 2018.

26. Martin S, Millán JR, Knight RT, Pasley BN. The use of intracranial recordings to decode human language: challenges and opportunities. Brain Lang. 2019;193:73–83. http://www.sciencedirect.com/science/article/pii/S0093934X15301243.

27. Martin S, Brunner P, Iturrate I, Millán JR, Schalk G, Knight RT, et al. Word pair classification during imagined speech using direct brain recordings. Sci Rep. 2016;6:srep25803. https://www.nature.com/articles/srep25803. Accessed 12 Sep 2017.

28. Ramsey NF, Salari E, Aarnoutse EJ, Vansteensel MJ, Bleichner MB, Freudenburg ZV. Decoding spoken phonemes from sensorimotor cortex with high-density ECoG grids. Neuroimage. 2018;180(Pt A):301–11.

29. Bouchard KE, Mesgarani N, Johnson K, Chang EF. Functional organization of human sensorimotor cortex for speech articulation. Nature. 2013;495(7441):327–32. http://www.nature.com/nature/journal/v495/n7441/full/nature11911.html?foxtrotcallback=true. Accessed 12 Sep 2017.

30. Glannon W. Ethical issues in neuroprosthetics. J Neural Eng. 2016;13(2):021002.

31. Birbaumer N, Cohen LG. Brain–computer interfaces: communication and restoration of movement in paralysis. J Physiol. 2007;579(3):621–36. http://onlinelibrary.wiley.com/doi/10.1113/jphysiol.2006.125633/abstract.
32. Kopsky DJ, Winninghoff Y, Winninghoff ACM, Stolwijk-Swüste JM. A novel spelling system for locked-in syndrome patients using only eye contact. Disabil Rehabil. 2014;36(20):1723–7. https://doi.org/10.3109/09638288.2013.866700. Accessed 12 Jun 2017.
33. Wolpaw JR, McFarland DJ, Neat GW, Forneris CA. An EEG-based brain-computer interface for cursor control. Electroencephalogr Clin Neurophysiol. 1991;78(3):252–9.
34. Bahdanau D, Bosc T, Jastrzębski S, Grefenstette E, Vincent P, Bengio Y. Learning to compute word embeddings on the fly. arXiv:170600286 [cs]. 2017. http://arxiv.org/abs/1706.00286. Accessed 7 Aug 2018.
35. Farahany NA. A neurological foundation for freedom. Stanford Technol L Rev. 2011;2011:4–15.
36. Baars B. How brain reveals mind neural studies support the fundamental role of conscious experience. J Conscious Stud. 2003;10(9–10):100–14. http://www.ingentaconnect.com/content/imp/jcs/2003/00000010/F0020009/art00008. Accessed 13 Jun 2017.
37. Tamburrini G. Brain to computer communication: ethical perspectives on interaction models. Neuroethics. 2009;2(3):137–49.
38. Rainey S. 'A steadying hand': ascribing speech acts to users of predictive speech assistive technologies. J Law Med. 2018;26(1):44–53.
39. Angrick M, Herff C, Mugler E, Tate MC, Slutzky MW, Krusienski DJ, et al. Speech synthesis from ECoG using densely connected 3D Convolutional Neural Networks: bioRxiv. 2018. http://biorxiv.org/lookup/doi/10.1101/478644. Accessed 25 Jan 2019.
40. Bengio Y, Ducharme R, Vincent P, Jauvin C. A neural probabilistic language model. J Mach Learn Res. 2003;3:1137–55.
41. Golosio B, Cangelosi A, Gamotina O, Masala GL. A cognitive neural architecture able to learn and communicate through natural language. PLoS One. 2015;10(11):e0140866. http://journals.plos.org/plosone/article?id=10.1371/journal.pone.0140866. Accessed 9 May 2018.
42. Mitchell RJ, Bishop JM, Low W. Using a genetic algorithm to find the rules of a neural network. Artificial Neural Nets and Genetic Algorithms. Springer, Vienna. 1993.
43. Doshi-Velez F, Kim B. Towards a rigorous science of interpretable machine learning. arXiv preprint arXiv:170208608. 2017.
44. Schooler JW, Engstler-Schooler TY. Verbal overshadowing of visual memories: some things are better left unsaid. Cogn Psychol. 1990;22(1):36–71.
45. Klaming L, Haselager P. Did my brain implant make me do it? Questions raised by DBS regarding psychological continuity, responsibility for action and mental competence. Neuroethics. 2013;6(3):527–39. https://link.springer.com/article/10.1007/s12152-010-9093-1. Accessed 4 May 2018.
46. Goering S, Klein E, Dougherty DD, Widge AS. Staying in the loop: relational agency and identity in next-generation DBS for psychiatry. AJOB Neurosci. 2017;8(2):59–70. https://doi.org/10.1080/21507740.2017.1320320. Accessed 13 Dec 2018.
47. Sharp D, Wasserman D. Deep brain stimulation, historicism, and moral responsibility. Neuroethics. 2016;9(2):173–85. https://link.springer.com/article/10.1007/s12152-016-9260-0. Accessed 30 May 2018.
48. Mesgarani N, Cheung C, Johnson K, Chang EF. Phonetic feature encoding in human superior temporal gyrus. Science. 2014;343(6174):1006–10. http://science.sciencemag.org/content/343/6174/1006. Accessed 12 Sep 2017.
49. Perfect TJ, Hunt LJ, Harris CM. Verbal overshadowing in voice recognition. Appl Cogn Psychol. 2002;16(8):973–80. https://onlinelibrary.wiley.com/doi/abs/10.1002/acp.920. Accessed 11 Jun 2018.
50. Vidaurre C, Blankertz B. Towards a cure for BCI illiteracy. Brain Topogr. 2010;23(2):194–8. https://doi.org/10.1007/s10548-009-0121-6. Accessed 19 Mar 2019.

The Emperor's New Clothes? Transparency and Trust in Machine Learning for Clinical Neuroscience

14

Georg Starke

Contents

Abstract

Machine learning (ML) constitutes the backbone of many applications of artificial intelligence. In the field of clinical neuroscience, applying ML to neuroimaging data promises wide-ranging advancements. Yet, such potential diagnostic and predictive tools pose new challenges with regard to old problems of transparency and trust. After all, the very design of many ML applications can preclude comprehensive explanations of its inner workings and impede accurate predictions about its future behavior, supposedly clashing with the ideal of transparency. It is often claimed that these shortcomings, inherent to many ML applications, are detrimental to their trustworthiness and thus hinder implementing new and potentially beneficial techniques. In this chapter, I will argue against beliefs that inextricably link transparency and trustworthiness. Drawing in particular on the framework of the British philosopher and bioethicist Onora O'Neill, I aim to show why, contrary to many intuitions, an obsession with transparency can be detrimental to tackling more fundamental ethical issues—and that hence

G. Starke (✉)
Institute for Biomedical Ethics, University of Basel, Basel, Switzerland
e-mail: georg.starke@unibas.ch

© Springer Nature Switzerland AG 2021
O. Friedrich et al. (eds.), *Clinical Neurotechnology meets Artificial Intelligence*,
Advances in Neuroethics, https://doi.org/10.1007/978-3-030-64590-8_14

transparency may not solve as many challenges for clinical ML applications as is usually assumed. I will conclude with a tentative suggestion on how to move forward from a practical point of view as to advance the trustworthiness of ML for clinical neuroscience.

14.1   Introduction

Mehr an Information und Kommunikation allein erhellt die Welt nicht.
Die Durchsichtigkeit macht auch nicht hellsichtig.[1]
Byung-Chul Han, Transparenzgesellschaft [2]

Machine learning (ML) constitutes the backbone of many applications of Artificial Intelligence (AI). In the field of clinical neuroscience, applying ML to data from neuroimaging promises wide-ranging possibilities, from assisting clinicians in diagnostic and prognostic exams to enabling the selection of an optimal pharmacological intervention [3–8]. Despite the increasing body of bioethical literature on the subject of ML in medicine [9–11], ethical discussions of ML with regard to neuroimaging data remain scarce [12, 13]. The aim of this chapter is to tackle this gap, focusing on the notion of transparency and its intricate relation to trust.

If we are to believe its proponents, transparency is key to solving the ethical challenges of clinically applied ML. Transparency is said to drive algorithmic fairness [14], guarantee patients' safety and enable informed consent [15]. According to some, it constitutes "the first step towards ethical and fair ML models" [16]. Unfortunately, the very design of many ML applications can preclude comprehensive explanations of its inner workings, thus posing particular challenges to an ideal of transparency [11, 17]. Black-box algorithms may prevent accurate predictions about future behavior, e.g., if the program continuously updates its inherent models based on newly available data. To many, such lack of scrutability of ML applications is of particular concern, as it can create gaps in responsibility for potential short fallings, which may also have legal implications [18, 19]. The use of poorly chosen or curated input data, for example, can result in errant or skewed output results, contributing to discriminatory or otherwise harmful practices [20, 21]. However, if the black-box program cannot be explained or understood, such errors may go unnoticed and evade remedy. Consequently, attributing responsibility to create clear, "transparent" patterns of accountability seems crucial for establishing a procedure's trustworthiness. In turn, trustworthiness may determine whether patients and physicians factually trust and thus embrace the clinical implementation

---

[1]Unfortunately, the English translation by Erik Butler cannot quite grasp the meaning of the German original: "More information and communication alone do not illuminate the world. Transparency also does not entail clairvoyance" [1].

of ML [22]. Hence, do we need to crack open the black boxes of ML applications in order to achieve transparency and render them trustworthy?

Placing so much burden on one scientific ideal certainly warrants scrutiny. Similar to the garments in Hans Christian Andersen's famous tale, which supposedly expose the viewers' own inadequacy, I will show why mere calls for transparency are too little to cloak the ethical challenges posed by applied ML in clinical neuroscience. Drawing on the writings of the philosopher and bioethicist Onora O'Neill, I will argue that transparency is not an all-purpose remedy for fostering trustworthiness and that an obsession with transparency can be detrimental to tackling more fundamental issues. To do so, I will discuss the ideal of transparency and its relation to trust in clinical ML procedures using neuroimaging data, in order to give a more practical demonstration of an abstract debate. This example may prove particularly fruitful since both transparency and neuroimaging share a common aim: to make things visible. Nevertheless, similar points regarding trust and transparency could be raised with regard to other clinical ML applications as well.

The structure of this chapter will be as follows: I will first provide a brief definition of ML and offer some examples of potential applications for clinical neuroscience. I will then outline why transparency is commonly thought to be an ethically important epistemic ideal and why it is ascribed prudential value to foster public trust in such new technological developments. In a third section, I will draw on O'Neill's philosophy and argue against beliefs that inextricably link transparency and trustworthiness. Finally, I will conclude with a tentative suggestion on how to move forward from an O'Neillian point of view and advance trust in beneficial uses of ML in neurotechnology by moving towards a form of "intelligent openness" [23].

14.2   Opportunities for Applied Machine Learning in Clinical Neuroscience

The notion of machine learning encompasses many different methods that constitute current state-of-the-art applications of artificial intelligence. An influential operational definition, which I will also use in this chapter, stems from the computer scientist Tom Mitchell, who characterized ML in terms of experience E, an area of tasks T, and performance measure P. Following Mitchell, ML describes a program "if its performance at tasks in T, as measured by P, improves with experience E" [24]. A classic application would be recognizing patterns, for instance across many different images. ML demarcates a narrower field than the broader term AI since the latter would also encompass the "holy grail" of AI research: generalized AI, capable of *any* intellectual task usually performed by humans, which for the moment remains in the domain of science fiction, such as the famous sentient computer HAL 9000 in Stanley Kubrick's *Space Odyssey*. At the same time, ML comprises a multitude of different computational approaches such as support vector machines (SVM) or artificial neural networks for deep learning (DL).

Neuroimaging, in turn, denotes as diverse an area as ML. Narrowly, it can be defined as "all techniques in which actual images of the brain are acquired" [25].

Such processes rely on vastly different acquisition techniques, such as computed tomography (CT), magnetic resonance imaging (MRI), positron emission tomography (PET), or maps derived from electroencephalography (EEG) or transcranial magnetic stimulation (TMS). Within each of these, many further important distinctions can be made, e.g., in MRI between structural and (task- or resting state) functional MRI, between different acquisition sequences such as spin echo or gradient echo and so forth. Comprehensive overviews over the different techniques are offered elsewhere [25], but it is important to note that depending on the imaging modality it is already the case today that different degrees of computational efforts and human agency are required to generate images, rendering the techniques differently close to ideals of mechanical objectivity [26].

Of course, applications that combine the two very diverse areas of ML and neuroimaging are highly heterogeneous themselves, adding yet another layer of complexity. Nevertheless, some broad distinctions can be made based on the different purposes for which ML-driven neuroimaging is employed. For the question at hand, the most fundamental distinction appears to be whether an application is intended (primarily) for research or for clinical purposes since this distinction shapes legal and ethical obligations involved in the process and may also determine the scope of the application. For example, let us suppose a project's only aim consists in contributing to a better understanding of pathologies underlying a certain disease or disorder. Let us further suppose, as a real-life example, that this project identifies different subtypes of schizophrenia based on distinct connectivity patterns by using ML on diffusion-weighted MRI scans. Critics may argue that such research could ultimately contribute to an unwarranted reification of psychiatric disorders [27] by creating a dubious classificatory system of supposed "natural kinds" [13]. But within such a research setting, patients would not be wronged by the tentative assignment to newly created, experimental subcategories of an already-diagnosed disorder.

However, where ML is applied to the clinic directly, stakes seem far higher. From identifying patients at risk of psychosis [28, 29] to predicting the course of multiple sclerosis [30], from prognostic tests for Alzheimer's dementia [31] to algorithms suggesting ideal psychopharmacologic drugs for Major Depressive Disorder [4, 32], future patients will certainly be treated based on recommendations by programs relying on ML and neuroimaging. If left unchecked, this could endanger the safety and well-being of patients, e.g., if the use of a diagnostic program provides the wrong diagnosis or misses a potentially life-threatening illness, leading in turn to wrong or delayed treatment. Potentially, recommendations may also be skewed by biased input data, reinforcing direct or indirect discriminatory practices [20]. Similar problems can arise with programs suggesting particular treatments. The shortcomings of IBM's Watson, recommending dangerous treatment strategies in oncology, may serve as a cautionary example here [33]. Many authors have thus argued that the development of such applications warrants special scrutiny to safeguard against unjustified systematic biases, to ascertain the clinical safety of their deployment and to establish clear chains of responsibility. Yet, given the vast disparities in potential uses and methods of ML for clinical neuroscience, it seems surprising that transparency is often treated as an all-purpose remedy for potential challenges posed by its implementation. A recent review of translational ML for psychiatric neuroimaging for example suggested that "[f]or use in clinical support systems, transparency is both necessary and sufficient for legal certification and patient safety" [34]. It thus seems worth shedding more light on this heavily burdened ideal.

14.3   The Ideal of Transparency

For a clear discussion of the ethical role of transparency, one firstly needs to note that the term itself encompasses two mutually exclusive meanings [15]. On the one hand, in computer science transparency commonly refers to a process's property of being *invisible* to the user. For example, if a common text-editing program is updated to make it run faster, its external, visible user-interface may not change at all, even though the underlying computational processes may be quite different in the new version. Still, such invisible change in the program's internals would be called "transparent" [15]. On the other hand, transparency also denotes a process's property of being *visible*, often in the combined term "algorithmic transparency." As an ideal, it describes programs that enable the user to *see* and scrutinize its internals, i.e., the underlying computational processes taking place [35]. In the ethical and legal arguments that concern us here, transparency usually refers to the second meaning only, and I will adhere to it in the following.

Unfortunately, clear definitions of transparency are hard to come by, and quite frequently transparency serves as "an empty signifier that can be filled by very different interpretations" [36]. In a minimal definition, transparency merely describes "putting content in the public domain" [37]. Such disclosure is thought to be commendable because it allows shedding light on something otherwise hidden, or as the former member of the US supreme court Louis Brandeis remarked: "Sunlight is said to be the best of disinfectants; electric light the most efficient policeman" (Brandeis, quoted in [38]). Related to this are positions which define transparency according to its opposition to concealment as the "lifting the veil of secrecy" ([39], quoted in [40]). In political contexts, this renders it often interchangeable with notions of openness and even finds tangible expression in architecture. The glass dome of the Berlin Reichstag building by Sir Norman Foster, promising an open, transparent government, may serve as a prime example [36]. Where transparency and secrecy are pitched against each other, they are often portrayed rather simplistically as a clash between good and evil [36].

Depending on the context, "transparency" is then adorned with a domain-specific epithet, from governmental transparency [41] and information transparency [15] to algorithmic transparency [35], to name but three prominent examples. Of course, the kind of transparency commonly invoked with regard to ML aligns closely with the last kind. However, as Albert Meijer has argued, an understanding of transparency rooted in modern discourse about information technology has become so pervasive that the two have become almost synonymous: "Modern transparency *is* computer-mediated transparency" ([24, 40], emphasis added). According to Meijer, three key components characterize this particular form of transparency. First, it is unidirectional in the sense that unlike in public assemblies, where information is mutually exchanged between various agents, information flows in one direction only. Second, computer-mediated transparency usually entails that the transferred information is taken out of its context, i.e., disconnected from its original theoretical and practical underpinnings. The reason for this lies in a third characteristic, namely that modern transparency is calculative and hence mostly requires quantifiable information: "computer-mediated forms of transparency reflect certain aspects of reality, namely those aspects that are being measured" [24, 40].

A crucial reason why many hold transparency, understood as "visibility contingent upon observation" ([42] quoted in [38]), to be key in resolving ethical challenges—such as the challenges of applied ML—lies in its supposed function of enabling other important ethical principles. Fittingly, Matteo Turilli and Luciano Floridi [15] have dubbed transparent access to certain kinds of information "pro-ethical," namely where disclosure of information has an impact on ethical principles. In the particular context of ML, transparency is thought to help establish accountability and promote fairness, by baring unjustified biases resulting in discriminatory differential treatment of salient social classes [17]. In doing so, transparency seems a vital condition for rendering clinically applied machine learning trustworthy.

14.4   Trust and Trustworthiness

Interest in different forms of trust has begotten immense corpora of academic literature throughout the past decades. Philosophers and social scientists alike have tackled the topic from different angles, motivated by the fact that it constitutes a central, but long-neglected phenomenon [43, 44]. The resulting definitions of trust differ largely, and so do its classifications, distinguishing between goodwill trust, competence trust, contractual trust, calculus-based trust, knowledge-based trust, and identification-based trust, to name but a few [45]. It suffices here to look at some of its general properties that are important to our inquiry and common to both interpersonal and institutional (or public) trust [46].[2] Two overlapping perspectives shape the debate: sociologists such as Georg Simmel, Niklas Luhmann, and Barbara Misztal are often primarily concerned with the *functions* of trust [44, 48, 49], while theorists who take a more philosophical approach such as Annette Baier, Russell Hardin, or Onora O'Neill aim for *definitory* clarifications [50–52].

Most accounts agree, from a conceptual point of view, that trust takes the form of a three-part relation: A trusts B regarding X, where X might range from particular actions, including speech acts, to general assumptions about B's behavior or character [43, 52]. Usually, A will only place her trust in B if she believes B to be

---

[2] In doing so, I will necessarily leave out many important facets of trust, especially conative or emotional components, which arguably are vital for full-fledged forms of trust. Some even hold that only non-cognitive trust should be regarded as trust in the fullest sense [47].

competent with regard to X and also to be *committed* to X.[3] For example, I will only entrust my neighbor with taking care of my flowers while on vacation if I take her to be capable of treating plants appropriately and am also optimistic about her actually looking after them. At the same time, I might not trust her to take care of my pet, knowing that she is highly averse to dogs. Note that this does not presuppose that she bears some form of goodwill towards me; she might well find me to be the most annoying person and still live up to my expectations, e.g., because she cares about her view onto my balcony or simply wants to have a good relationship with her neighbours.[4] Analogous claims can be made about trust in institutions, where the parties may also not act out of emotional attachment but rather have an interest in future interactions with the truster. As Russell Hardin put it: "I trust you because I think it is in your interest to take my interests in the relevant matter seriously" [52].

While exact conceptualizations of trust remain challenging, the phenomenon's central functions seem clearer. As an expectation in specific situations of uncertainty, trust serves as a means to reduce social complexity in absence of certainty [44] and thus constitutes "a solution for specific problems of risk" [55]. After all, trust is only necessary where one lacks comprehensive knowledge or the capacity to exercise full control and hence needs, to some extent, to rely on the actions of others. To stay with the previous example, I would not need to trust my neighbor if I could either water my flowers myself or had means to fully determine her actions. As a means to deal with risk, trusting itself remains a "risky investment" [56] and renders us vulnerable to the actions of others. Annette Baier even uses the property of being open to betrayal as a core characteristic of trust, to distinguish it, e.g., from reliance [43].

Clinical applications of ML for decision-making in medicine doubtlessly entail risks and make patients vulnerable to failure on multiple levels. But is trust really the correct way of dealing with such risks? After all, it is crucial not to trust blindly and only place trust in trustworthy agents. "Not to trust rashly is the nerves and joints of wisdom," Cicero advised his brother [57]. For ethical debates with a view on practical implementation, evaluating the trustworthiness of an agent, institution, or procedure may thus be more pressing than an abstract debate about the phenomenon of trust [58]. While there is much debate about properties that warrant trustworthiness, a few points stand out as uncontroversially detrimental, as they follow from the previously discussed expectations of the trustees to be capable and committed to act in our interest. We do not want to trust people with tasks they cannot possibly achieve [50, 59] and even less if they invite our trust by lying or withholding critical information in a deceiving manner [54]. Additionally, we expect trustees to act on our interests, so it can undermine agents' trustworthiness if they have important competing interests that are opposed to our own. Importantly, anyone

---

[3] However, exceptions exist, for example parents placing trust in their children as a means of education, even when they are convinced that the children will fall short of their trust [53].

[4] I am following O'Neill here [54], who disagrees on this point with Annette Baier [43].

abusing their power against us would also be very untrustworthy, for we certainly do not want to increase this power further by placing unjustified trust in them.

What could this mean with regard to clinically applied ML? Many authors agree that as a remedy for achieving trustworthiness of such new methods we need to turn to transparency. The assumed mechanism seems to be that transparency increases knowledge about a procedure and thus renders it more trustworthy insofar as it decreases the uncertainty necessarily involved in trusting. In their recent guidelines, the EU Commission's High Level Expert Group on Artificial Intelligence explicitly names transparency as one of its seven key requirements for trustworthy AI. Similarly, and with particular regard to medical ML (MLm), Vayena et al. state that a "lack of transparency can preclude the mechanistic interpretation of MLm-based assessments and, in turn, reduce their trustworthiness" [11]. Certainly, trust-worthiness can be enhanced in certain instances by tangible forms of (algorithmic) transparency. However, it is far from clear that the relation between transparency and trust is as straightforward as is commonly assumed.

14.5   The Paradox Relation of Trust and Transparency

Does transparency beget trust—or are matters more complicated? The British philosopher and bioethicist Onora O'Neill has long addressed this question from several angles. To tackle the intricacies of transparency of clinically applied ML, her framework appears particularly well suited since she developed her account in the very context of biomedical ethics. Three of her works stand out as highly instructive: her seminal *Autonomy and Trust in Bioethics*, the BBC Reith Lectures *A Question of Trust,* and *Rethinking Informed Consent in Bioethics,* co-authored with Neil Manson. From these three works, three key ideas can be distilled that have not yet been applied to clinical ML but can guide further discussions on its ethical dimension.

First, in her *Autonomy and Trust in Bioethics*, O'Neill highlights that factually, transparency does not simply beget trust. To do so, she extensively discusses trust in the so-called "risk society" [54], referring to the work of the German sociologist Ulrich Beck [60]. From a sociological perspective, risk societies are characterized by increased public fears and anxieties about hidden risks of increasingly complex social and technological practices. Frequently, these "focus particularly on hazards introduced (or supposedly introduced) by high-tech medicine" [54]. Importantly though, risk societies are *not necessarily* characterized by factually increased risks in all areas of society.[5] In fact, risk societies are often wealthier, healthier, and less secretive than their historical precursors. Still, due to a changed *perception* of risks, people in these societies tend to be more reluctant to placing trust. The medical domain provides a particular striking example for such erosion of trust. In the United States, public trust in the medical profession declined sharply from 73%

---

[5] Of course, this is not to deny the grave risks created by modern societies in specific areas, e.g., newly introduced environmental risks [60].

reported in 1966 to 34% in 2012 [61]. Yet, provisions to increase transparency and foster the autonomous and informed decision-making of patients have undoubtedly greatly increased since the 1960s, when paternalist doctor–patient relationships were still the norm.[6] How may one explain such observations, undermining a supposed close link between transparency and trust? One tentative explanation could be that if people are generally distrustful of technological advances, transparency may have limited influence on such a societal phenomenon since people may still remain distrustful against any transparently disclosed information. O'Neill calls such a public mood, potentially diminishing the impact of transparency of trust, a "culture of suspicion" [51]. In an imaginary society of trust though, things may be just the opposite. Due to higher generalized trust, people may be more prone to placing trust in transparently disclosed information about medical technologies. In other words, we find the paradox that effective disclosure, which could increase a procedure's trustworthiness, seemingly presupposes trust.[7]

In her *Reith Lectures*, O'Neill discusses this entangled relation in more detail. In line with Meijer's previously discussed arguments about modern transparency being computer-mediated, also for O'Neill, transparency constitutes the "new ideal of the information age." [51] In political contexts, increasing transparency often seems coextensive with improving structures of accountability, supposedly fostering trust in public and professional institutions. However, accountability does not aim at establishing relations of trust but rather at minimizing risks, ideally improving a trustee's trustworthiness.[8] However, mere transparency in the sense of "putting content in the public domain" [37] may not increase trustworthiness, if it does not aim at increasing the truster's knowledge, e.g., because it is accessible but incomprehensible. Thus, while increased transparency may often have beneficial effects, it is too little for establishing a trustee's trustworthiness [37]. At the same time, focusing solely on transparency runs danger of marginalizing other, more basic obligations [35]. In particular, O'Neill argues, these comprise the true enemy of trust, which is neither secrecy nor a lack of information, but willful deception. In fact, such deception can come in the very guise of supposed transparency, namely when agents hide behind a rallying cry of transparency to engage in information dumping, burying future trusters under a heap of unsorted or misleading data, confusing the addressee, and obscuring the actually crucial information. Of course, this is not to say that some forms of transparency may not render a process trustworthy. Full and transparent disclosure of information can, in some instances and particularly after previous cases of deception, improve the trustworthiness of agents and institutions. Yet, O'Neill argues, an obsession with transparency can also undermine trust: like plants, trust does not flourish when constantly uprooted [54].

---

[6] It may be worth recalling here that the principlist framework by Beauchamp and Childress, stressing respect for autonomy as a fundamental principle of medical practice, was only published in 1979.

[7] Annette Baier has made a similar point, stressing that "trust [is] a response to perceived trustworthiness, [...] but it is equally true that trustworthiness is, to some degree, a response to trust" [62].

[8] Of course, this is not to say that trust is a mere matter of risk calculation.

In *Rethinking Informed Consent in Bioethics,* O'Neill and Manson provide a practical demonstration of these considerations and show how effective disclosure requires trust, with specific regard to informed consent. Following Willard Van Orman Quine's distinction of referential transparency and opacity, they stress that informed consent is referentially opaque [63]. Their general idea is that if A consents to B doing p, she does thereby not consent to q, even if the propositions q and p are logically equivalent. Drawing on an example from Ruth Faden and Tom Beauchamp [64], O'Neill and Manson highlight that a person may consent to taking lysergic acid diethylamide as part of a study. Yet, the same person would possibly not consent to the very same procedure if she was asked to take LSD—even though, of course, the two are identical. On a more applied level, this abstract debate finds its expression in the requirements of *valid* informed consent, which entails that the patient or research subject actually understands the proposition to which she consents. Researchers or physicians asking for consent hence need to consider the level of knowledge and the beliefs of the consenters, who in turn may place their trust in them that the information provided is correct, comprehensive, and adequately understandable. In other words, mere transparency is not sufficient since effective transference of information requires taking into account the audience's needs and interests to make sure that the information and its implications are properly understood. To stay with the example, putting LSD's technical name in a consent sheet instead of providing the commonly known abbreviation could thus very well amount to willful deception—as could the mere publication of an incomprehensible source code for a medical ML application.

14.6 Trust and Transparency of Applied ML for Neuroimaging

If O'Neill is right and trust is required for effective communication, it will prove impossible to avoid. Certainly, this holds true within medicine, inherently shaped by unknown and unknowable risks. Accordingly, many authors have addressed issues of trust in the medical domain in past years and both interpersonal and public trust have received plenty of attention from medical ethicists [54, 65]. For example, Gille et al. have a model of public trust in medicine which draws on Habermas' Theory of Communicative Action and acknowledges the centrality of trust as a necessary condition between communicating parties [66]. However, the necessity of trust goes far beyond communication and constitutes a necessity of everyday life, or as Luhmann put it: "a complete absence of trust would prevent [one] even from getting up in the morning" [56].

It should hence not come as a surprise if new advances in medical ML, whether for diagnostic, therapeutic, or prognostic purposes, will require trust for their successful implementation. In fact, trust seems necessary for the acceptance of ML-assisted decision-making not only by patients, but also by physicians, nurses, and other health care professionals. However, as this chapter aimed to show, mere transparency cannot guarantee trustworthiness. So what could this mean practically if applications of ML drawing on neuroimaging data gain ground? As a first step, acknowledging that mere unidirectional disclosure is not a remedy for all potential problems may create space to make more fundamental notions central. Making the source code publicly available is certainly commendable in research contexts, but in itself too little to render a program trustworthy for patients and physicians. Secondly, this opening would need to be filled by principles that are more substantial. In her most recent book, O'Neill sketches such an alternative model which she calls "intelligent openness":

> Scientific communication [...] requires not mere transparency, but 'intelligent openness' that ensures that communication is in principle accessible, intelligible and assessable for all others, so fully open to their check and challenge [23].

The recent EU guidelines for trustworthy AI seem to provide a step in this direction, as they link transparency explicitly to communication. With specific regard to medical ML, programs developed to be understandable by patients and physicians may be much better suited to build (on) trusting relationships, as they aim at successful communication. For example, so-called Influence Style Explanations, which estimate the impact of distinct inputs on a given output, already in use for non-medical recommender systems, could be implemented for applications of medical ML as well [14]. In clinical contexts, this could, e.g., take the form of a weighted list of parameters taken into account for a program's suggestion. In neuroimaging, understandable ML programs could further provide visualizations, which have long been a focus for examining and explaining ML [67]. Such applications could, e.g., highlight visual aspects in the data taken to be salient, such as newly acquired white matter lesions for MS progress reports.

Two twists need to be noted in discussions of medical ML for neuroimaging. First, a debate regarding the transparency of ML may stress that already the input itself, i.e., neuroimaging visualizations, should not be considered belief-transparent, mechanically objective representations of the outward world. Given the underlying complex interactions between humans and machines, most imaging modalities cannot be ascribed the same epistemic value as photographs. As Adina Roskies has noted with regard to fMRI:

> We do not 'see through' the visual properties of neuroimages to the visual properties of their subjects; we do not understand the causal and counterfactual relationships between the images and the data they represent to the same extent that we understand them with photography [68].

Reconsidering transparency of ML applications may hence inadvertently revitalize existing discourses about the epistemic status of neuroimaging modalities and draw further attention to the training required for their proper understanding [69]. Second, as Stephen John has argued with regard to science communication and climate change, expectations of transparency have become so engrained in our societal discourses that transparency can be seen as a fundamental principle of a "folk philosophy of science" [70]. Hence, while transparency may in fact not beget trust, an open denial of transparency could damage trust by breaking (arguably unrealistic) expectations. Again, a stronger focus on a more intelligent form of openness oriented towards communication may offer a way out of this dilemma by providing a related, yet more substantial rallying cry.

Doubting the primacy of transparency will certainly prove challenging. The German-Korean philosopher Byung-Chul Han, a long-standing critic of the "transparency society," has warned against possible resistance: "The imperative of transparency is suspicious of everything that does not submit to visibility" [1, 2]. Still, such change of focus may be vital to increasing public trust in clinically applied ML. As Annette Baier has noted, "trust comes in webs, not in single strands" [50]. Weaving such webs anew to accommodate for new medical technologies thus seems to pose a particular challenge—but one worth pursuing.

References

1. Han B-C. The transparency society. Stanford: Stanford University Press; 2015.
2. Han B-C. Transparenzgesellschaft. Berlin: Matthes & Seitz; 2012.
3. Huys QJ, Maia TV, Frank MJ. Computational psychiatry as a bridge from neuroscience to clinical applications. Nat Neurosci. 2016;19(3):404–13.
4. Webb CA, Trivedi MH, Cohen ZD, Dillon DG, Fournier JC, Goer F, et al. Personalized prediction of antidepressant v. placebo response: evidence from the EMBARC study. Psychol Med. 2018;49(7):1118–27.
5. Janssen RJ, Mourao-Miranda J, Schnack HG. Making individual prognoses in psychiatry using neuroimaging and machine learning. Biol Psychiatry Cogn Neurosci Neuroimaging. 2018;3(9):798–808.
6. Dwyer DB, Cabral C, Kambeitz-Ilankovic L, Sanfelici R, Kambeitz J, Calhoun V, et al. Brain subtyping enhances the neuroanatomical discrimination of schizophrenia. Schizophr Bull. 2018;44(5):1060–9.
7. Brodersen KH, Deserno L, Schlagenhauf F, Lin Z, Penny WD, Buhmann JM, et al. Dissecting psychiatric spectrum disorders by generative embedding. Neuroimage Clin. 2014;4:98–111.
8. Xiao Y, Yan Z, Zhao Y, Tao B, Sun H, Li F, et al. Support vector machine-based classification of first episode drug-naive schizophrenia patients and healthy controls using structural MRI. Schizophr Res. 2019;214:11–7.
9. Darcy AM, Louie AK, Roberts LW. Machine learning and the profession of medicine. JAMA. 2016;315(6):551–2.
10. Char DS, Shah NH, Magnus D. Implementing machine learning in health care—addressing ethical challenges. N Engl J Med. 2018;378(11):981–3.
11. Vayena E, Blasimme A, Cohen IG. Machine learning in medicine: addressing ethical challenges. PLoS Med. 2018;15(11):e1002689.
12. Martinez-Martin N, Dunn LB, Roberts LW. Is it ethical to use prognostic estimates from machine learning to treat psychosis? AMA J Ethics. 2018;20(9):E804–11.
13. Bzdok D, Meyer-Lindenberg A. Machine learning for precision psychiatry: opportunities and challenges. Biol Psychiatry Cogn Neurosci Neuroimaging. 2018;3(3):223–30.
14. Abdollahi B, Nasraoui O. Transparency in fair machine learning: the case of explainable recommender systems. In: Zhou J, Chen F, editors. Human and machine learning. Basel: Springer International Publishing; 2018. p. 21–35.
15. Turilli M, Floridi L. The ethics of information transparency. Ethics Inf Technol. 2009;11(2):105–12.
16. Zhou J, Chen F. Human and machine learning: visible, explainable, trustworthy and transparent. Basel: Springer International Publishing; 2018.

17. Kroll JA, Huey J, Barocas S, Felten EW, Reidenberg JR, Robinson DG, et al. Accountable algorithms. U Penn Law Rev. 2017;165(3):633–705.
18. Bublitz C, Wolkenstein A, Jox RJ, Friedrich O. Legal liabilities of BCI-users: responsibility gaps at the intersection of mind and machine? Int J Law Psychiatry. 2019;65:101399.
19. Matthias A. The responsibility gap: ascribing responsibility for the actions of learning automata. Ethics Inf Technol. 2004;6(3):175–83.
20. Favaretto M, De Clercq E, Elger BS. Big data and discrimination: perils, promises and solutions. A systematic review. J Big Data. 2019;6(1):12.
21. Cohen IG, Amarasingham R, Shah A, Xie B, Lo B. The legal and ethical concerns that arise from using complex predictive analytics in health care. Health Aff. 2014;33(7):1139–47.
22. Schnall R, Higgins T, Brown W, Carballo-Dieguez A, Bakken S. Trust, perceived risk, perceived ease of use and perceived usefulness as factors related to mHealth technology use. Stud Health Technol. 2015;216:467–71.
23. O'Neill O. From principles to practice: normativity and judgement in ethics and politics. Cambridge: Cambridge University Press; 2018.
24. Mitchell TM. Machine learning. New York: McGraw-Hill; 1997.
25. Kellmeyer P. Ethical and legal implications of the methodological crisis in neuroimaging. Camb Q Healthc Ethics. 2017;26(4):530–54.
26. Daston L, Galison P. Objectivity. New York: Zone Books; 2007.
27. Hyman SE. The diagnosis of mental disorders: the problem of reification. Annu Rev Clin Psychol. 2010;6:155–79.
28. Ramyead A, Studerus E, Kometer M, Uttinger M, Gschwandtner U, Fuhr P, et al. Prediction of psychosis using neural oscillations and machine learning in neuroleptic-naive at-risk patients. World J Biol Psychiatry. 2016;17(4):285–95.
29. Koutsouleris N, Riecher-Rossler A, Meisenzahl EM, Smieskova R, Studerus E, Kambeitz-Ilankovic L, et al. Detecting the psychosis prodrome across high-risk populations using neuroanatomical biomarkers. Schizophr Bull. 2015;41(2):471–82.
30. Zhao Y, Healy BC, Rotstein D, Guttmann CR, Bakshi R, Weiner HL, et al. Exploration of machine learning techniques in predicting multiple sclerosis disease course. PLoS One. 2017;12(4):e0174866.
31. Dallora AL, Eivazzadeh S, Mendes E, Berglund J, Anderberg P. Machine learning and microsimulation techniques on the prognosis of dementia: a systematic literature review. PLoS One. 2017;12(6):e0179804.
32. Chekroud AM, Zotti RJ, Shehzad Z, Gueorguieva R, Johnson MK, Trivedi MH, et al. Cross-trial prediction of treatment outcome in depression: a machine learning approach. Lancet Psychiatry. 2016;3(3):243–50.
33. Ross C, Swetlitz I. IBM's Watson supercomputer recommended 'unsafe and incorrect' cancer treatments, internal documents show. 2018. https://www.statnews.com/2018/07/25/ibm-watson-recommended-unsafe-incorrect-treatments/.
34. Walter M, Alizadeh S, Jamalabadi H, Lueken U, Dannlowski U, Walter H, et al. Translational machine learning for psychiatric neuroimaging. Prog Neuro-Psychopharmacol Biol Psychiatry. 2019;91:113–21.
35. Desai D, Kroll J. Trust but verify: a guide to algorithms and the law. Harv J Law Technol. 2017;31(1):1–64.
36. Worthy B. Transparency. In: Nerlich BH, Sarah H, Raman S, Smith A, editors. Science and the politics of openness: here be monsters. Manchester: Manchester University Press; 2018. p. 23–32.
37. O'Neill O, Bardrick J. Trust, trustworthiness and transparency. European Foundation Centre: Brussels; 2015.
38. Hansen HK, Flyverbom M. The politics of transparency and the calibration of knowledge in the digital age. Organization. 2015;22(6):872–89.
39. Davis J. Access to and transmission of information: position of the media. In: Deckmyn V, Thomson I, editors. Openness and transparency in the European Union. Maastricht: European Institute of Public Administration; 1998. p. 121–6.

40. Meijer A. Understanding modern transparency. Int Rev Adm Sci. 2009;75(2):255–69.
41. Meijer A. Understanding the complex dynamics of transparency. Public Admin Rev. 2013;73(3):429–39.
42. Brighenti A. Visibility—a category for the social sciences. Curr Sociol. 2007;55(3):323–42.
43. Baier A. Trust and antitrust. Ethics. 1986;96(2):231–60.
44. Luhmann N. Vertrauen; ein Mechanismus der Reduktion sozialer Komplexität. Stuttgart: F. Enke; 1968.
45. Bachmann R. Trust, power and control in trans-organizational relations. Organ Stud. 2001;22(2):337–65.
46. Townley C, Garfield JL. Public trust. In: Mäkelä P, Townley C, editors. Trust: analytic and applied perspectives. Amsterdam: Rodopi; 2013. p. 95–108.
47. Becker LC. Trust as noncognitive security about motives. Ethics. 1996;107(1):43–61.
48. Möllering G. The nature of trust: from Georg Simmel to a theory of expectation, interpretation and suspension. Sociology. 2001;35(2):403–20.
49. Misztal BA. Trust in modern societies : the search for the bases of social order. Cambridge: Polity Press; 1996.
50. Baier A. Trust. In: The tanner lectures on human values. Salt Lake City: University of Utah Press; 1992.
51. O'Neill O. A question of trust. The BBC Reith Lectures 2002. Cambridge: Cambridge University Press; 2002.
52. Hardin R. Trust and trustworthiness. New York: Russell Sage Foundation; 2002.
53. McGeer V. Trust, hope and empowerment. Australas J Philos. 2008;86(2):237–54.
54. O'Neill O. Autonomy and trust in bioethics. Cambridge: Cambridge University Press; 2002.
55. Luhmann N. Familiarity, confidence, trust: problems and alternatives. In: Trust: making and breaking cooperative relations, vol 6; 2000. p. 94–107.
56. Luhmann N. Trust and power. English ed. Chichester: Wiley; 1979.
57. Cicero M. Commentariolum Petitionis. In: Watt W, editor. M Tulli Ciceronis Epistulae. III. Oxford: Oxford University Press; 1963.
58. O'Neill O. Trust before trustworthiness? In: Archard D, Deveaux M, Manson NC, Weinstock D, editors. Reading Onora O'Neill. Oxford: Routledge; 2013. p. 237–8.
59. Scanlon T. Promises and practices. Philos Public Aff. 1990;19:199–226.
60. Beck U. Risk society: towards a new modernity. London: Sage Publications; 1992.
61. Blendon RJ, Benson JM, Hero JO. Public trust in physicians—U.S. medicine in international perspective. N Engl J Med. 2014;371(17):1570–2.
62. Baier A. What is trust? In: Archard D, Deveaux M, Manson NC, Weinstock D, editors. Reading Onora O'Neill. Oxford: Routledge; 2013:175–85.
63. Quine WVO. From a logical point of view: 9 logico-philosophical essays. Cambridge: Harvard University Press; 1980.
64. Faden RR, Beauchamp TL, King NMP. A history and theory of informed consent. Oxford: Oxford University Press; 1986.
65. Gille F, Smith S, Mays N. Why public trust in health care systems matters and deserves greater research attention. J Health Serv Res Policy. 2015;20(1):62–4.
66. Gille F, Smith S, Mays N. Towards a broader conceptualisation of 'public trust' in the health care system. Soc Theory Health. 2017;15(1):25–43.
67. Zhou J, Chen F. 2D transparency space—bring domain users and machine learning experts together. Human and machine learning. Basel: Springer International Publishing; 2018. p. 3–19.
68. Roskies AL. Are neuroimages like photographs of the brain? Philos Sci. 2007;74(5):860–72.
69. Racine E, Bar-Ilan O, Illes J. fMRI in the public eye. Nat Rev Neurosci. 2005;6(2):159–64.
70. John S. Epistemic trust and the ethics of science communication: against transparency, openness, sincerity and honesty. Soc Epistemol. 2018;32(2):75–87.

The Security and Military Implications of Neurotechnology and Artificial Intelligence

15

Jean-Marc Rickli and Marcello Ienca

Contents

Abstract

This chapter aims at taking stock of technological advances in artificial intelligence (AI) and neurotechnology and looks at the security and military implications of these technologies in light of their current capabilities. AI and neurotechnology hold a great transformative potential due to their ability to read,

J.-M. Rickli
Geneva Centre for Security Policy (GCSP), Geneva, Switzerland
e-mail: j.rickli@gcsp.ch

M. Ienca (✉)
Department of Health Sciences and Technology (D-HEST), Swiss Federal Institute of Technology, ETH Zurich, Zurich, Switzerland
e-mail: marcello.ienca@hest.ethz.ch

© Springer Nature Switzerland AG 2021
O. Friedrich et al. (eds.), *Clinical Neurotechnology meets Artificial Intelligence*,
Advances in Neuroethics, https://doi.org/10.1007/978-3-030-64590-8_15

modify, simulate and amplify human cognition in a variety of domains and in response to a variety of cognitive and analytical tasks. Furthermore, both technologies are rapidly proliferating outside traditional supervised settings (e.g. the clinics and academic research) onto multiple and unsupervised domains, a phenomenon that can be labelled "horizontal proliferation". Among these domains, their co-optation into the military sector and subsequent weaponization are of particular concern from an international security perspective. For each technological category, five security-relevant issues are discussed: data bias and accountability, manipulation, social control, weaponization and democratization of access. We argue that, in light of their disruptive potential and rapid proliferation, both neurotechnology and artificial intelligence urge global governance responses that deal with their accessibility, their proliferation, their dual-use nature including how easily these technologies can be repurposed and obviously the ethics and values that should accompany the development and use of these technologies. These responses should be inclusive and comprise all the different stakeholders (governments, private sector, scientific community, civil society and tech companies) and be very versatile as these technologies and applications evolve rapidly.

15.1 Introduction

Cognition is a major driver of complex information processing, knowledge acquisition and adaptive behaviour in both biological organisms and artificial systems. In the last century, parallel advances in the mapping and functional understanding of the human brain, on the one hand, and the processing of information in artificial systems (e.g. computers), on the other hand, have led to the development of a variety of technologies that assist, augment or simulate human cognitive processes or that can be used to achieve cognitive aims. These technologies, sometimes referred to using the umbrella term "cognitive technology" [1, 2], can be classified into two main categories: (a) technologies that monitor, assist or enhance cognitive processes in biological organisms—human beings included—and (b) technologies that simulate (aspects of) natural cognitive processes through artificial systems. The first category, commonly referred to as neurotechnology, encompasses devices that interface biological nervous systems to monitor, assist or enhance the cognitive processes executed by those systems. The second category, commonly referred to as artificial intelligence (AI) (or cognitive computing), encompasses systems and devices that artificially simulate cognitive functions typically executed by biological nervous systems—especially human brains—such as learning, planning, reasoning and perceiving the environment. In the last decade, these two domains have increasingly converged due to a twofold trend. First, artificially intelligent features have increasingly been embedded in neurotechnologies in order to better extract, classify and decode neural signals. Second, following trends such as neuromorphic artificial intelligence, artificial cognitive systems have been inspired by the study of biological neural systems.

Due to their disruptive potential, both families of cognitive technology raise security concerns, especially due to potential military implications. These implications are associated with three shared characteristics of both types of cognitive technology, namely: (a) proliferation outside supervised research domains, (b) re-purposing for military aims and (c) highly transformative, even disruptive, potential. Physicist Stephen Hawking famously said a few months before he passed away that,

> success in creating effective AI could be the biggest event in the history of our civilization, or the worst. We just don't know. So we cannot know if we will be infinitely helped by AI, or ignored by it and sidelined or conceivably destroyed by it (quoted in [3]).

In recent years, similar predictions have been made by other prominent figures such as philosopher Nick Bostrom and entrepreneur Elon Musk, who both raised the prospects that technologies related to AI might turn bad. Similarly, lieutenant colonel of the United States Air Force Brian E. Moore has predicted that neurotechnology, especially brain-computer interfacing, "has the potential to revolutionize military dominance much the same way nuclear weapons have done" [4]. A fierce debate pitting proponents and adversaries of cognitive technology ensued. Though this debate is very often characterized by exaggerations, hyperboles or even fear-mongering statements about the either utopian or dystopian consequences of AI and neurotechnology, it has the merit to raise public awareness about the security implications of these emerging technologies.

This chapter aims at taking stock of technological advances in artificial intelligence and neurotechnology and looks at the security and military implications of these technologies in light of their current capabilities. For each technological category, five security-relevant issues are discussed: data bias and accountability, manipulation, social control, weaponization and democratization of access.

15.2   The Security Implications of Artificial Intelligence

There is no universally agreed upon definition of artificial intelligence. As noted by the group of researchers at the University of Helsinki, AI is a scientific discipline, meaning that AI is a "collection of concepts, problems and methods for solving them" [5]. Nonetheless, the definition provided by the independent high-level expert group on artificial intelligence of the European Commission is a good starting point. Thus, artificial intelligence systems are

> "software (and possibly also hardware) systems designed by humans that, given a complex goal, act in the physical or digital dimension by perceiving their environment through data acquisition, interpreting the collected structured or unstructured data, reasoning on the knowledge, or processing the information, derived from this data and deciding the best action(s) to take to achieve the given goal. AI systems can either use symbolic rules or learn

a numeric model, and they can also adapt their behaviour by analysing how the environment is affected by their previous actions" [6].

Artificial intelligence methods enable "machines to learn from experience, adjust to new inputs and perform human-like tasks" [7]. Nowadays, it is hard to come across artificial intelligence without encountering the words "machine learning" (ML) which is a subset of AI. Essentially, ML refers to the development of algorithms which progressively improve performance on a specific task by making and testing predictions on data without being explicitly programmed. ML provides computers with the ability to use data to teach themselves, instead of via humans who program the machine.

There are two different "types" or categories of AI, known as "narrow" or "weak" AI and "general" or "strong" AI [8]. The distinction here is between machines that can perform and outperform humans in one specific task, and machines that might be able to adapt to any tasks. Today we are good, and getting better, at "narrow" AI, but are still decades away from creating machines which can perform the wide array of human-like tasks of "general" AI. In a recent survey of AI experts, the median timeframe predicted for the achievement of artificial general intelligence (AGI) is 45 years from now [9]. If, and once AGI is reached some posit that then AI shall be very rapidly developed into superintelligence surpassing any human intelligence [10]. In this chapter, we are considering current state of AI, that is narrow AI. Narrow AI pulls information from one specific dataset; it is programmed to perform a single task and does not perform outside of that single task which it was designed to perform. Algorithms relying on narrow AI include those used by Google Translate, spam-filtering systems, facial recognition technology and algorithms designed to learn and play video games, for instance.

The past few years have seen major progress in the development of AI. This is not only due to the vast improvement of the algorithms and techniques used, such as deep learning or machine learning, but also due to the incredible computer capacity that is now available and the vast amount of data that can be used to train the algorithms better than ever before. Thus, Google Deepmind, through its Alpha-class algorithms, achieved superhuman capabilities at the games of chess, shogi and Go, competed at the same level as the best players of the video game Starcraft II and won a global competition based on folding proteins [11–13]. Another algorithm, Libratus, developed by Carnegie Mellon University, defeated the best Texas Hold 'Em Poker players in January 2017 [14]. In February 2019, the most advanced Natural Language processing (NLP) algorithm was developed by Open AI, a leading AI research organization based in San Francisco. NLP is a subcategory of artificial intelligence which focuses on training computers to understand and process human language. This is a particularly difficult strand of AI, as computers do not have the same intuitive understanding of human languages; computers cannot "read between the lines" and understand implied meaning. The Open AI algorithm was trained to predict the next word, given all the previous words within a text [15]. The result has been the ability to generate lengthy text samples of unprecedented quality based on an input [16].

The transformative nature of AI offers fantastic prospects for improving human life in every domain. Thus, algorithms have surpassed humans at image recognition, which has had positive implications in the medical imagery domain, for instance. Algorithms are now much better at reading MRI, scans or X-rays than doctors are, therefore also reducing the risk of mistakes [17]. However, this technology also entails potential risks related to their misuse or malevolent use. The following sections will deal with the issues of data bias and accountability, manipulation especially for political purposes, social control, military applications and democratization of access.

15.3 Data Bias and Accountability

As AI is highly dependent on the data that it is fed with, biased data will lead to biased results. An experiment at Massachusetts Institute of Technology (MIT) which fed an algorithm with data only depicting crime scenes and death, led the algorithm to interpret any picture as death-related. The researchers called this algorithm "Norman", the world's first psychopath AI [18]. This experiment convincingly demonstrates the crucial importance of the quality of data required to train algorithms and the consequent inherent problem of biases in artificial intelligence.

When applied to real-world problems, the use of such technology, not being entirely aware of real-world subtleties, can entail moral and ethical problems. This is true, for example, for the criminal justice system. In 2016, ProPublica released an investigation into a machine learning system used by some courts in the USA [19]. The system was used to predict which individuals would be more susceptible to commit another crime after their release. It was observed that a system originally intended to operate free of human bias, only perpetuated this bias on a wider scale [20]. Indeed, as it was fed with historical criminal data from a criminal justice system that is historically biased against African American individuals, the system rated black individuals more negatively than white individuals to the point that the predictive algorithm was twice as likely to incorrectly classify black defendants as being at a higher risk than whites. In this sense, the results represented an automation of bias. We can see here a clear ethical conundrum. This led the major American tech companies to regroup in the consortium "Partnership on AI" to speak out against the use of algorithms for jailing people [21].

Furthermore, the results yielded by an AI powered algorithm are by definition not transparent and explainable. This is called the black-box problem of AI [22]. Because of its complex mathematical and probabilistic operations, the accountability of the machine learning process is very difficult to guarantee. Indeed, once fed with certain inputs, it is very complex to understand how the algorithm goes about producing the outputs. This impedes the understanding of "why" an algorithm has come to a certain conclusion. If AI is to have an increasingly influential role in the world and control greater parts of our lives, it is essential that they are accountable because people and society will want to know "why" algorithms make certain decisions that determine access to loans, recommend a medical treatment, or identify national security threats. Being unable to answer these questions might reduce the overall trust in these systems and therefore hinder their adoption [23].

15.4 Manipulations

The 2016 US presidential election can be seen as a turning point in the history of political manipulation. A private company named "Cambridge Analytic" was involved in a disinformation campaign to sway political vote in favour of Republican candidate Donald Trump. Cambridge Analytica did this by targeting voters based on their personal data generated on social media and other digital platforms [24].

Now picture the same process with an incredibly accurate AI, capable of automating the creation of fake and targeted content and flooding the web so that everybody could potentially receive personalized advertising and information that only reinforce held beliefs. This would raise enormous ethical and political concerns as it would undermine democratic processes by enabling malicious actors to stir political debate and dilute the truth.

The development of generative adversarial networks (GANs)—which are algorithms pitting neural networks against each other—has made it possible to manipulate data to a level unseen before, notably through deepfake which is a technique that superimposes images and videos onto other source images or videos. Deepfake pornography surfaced on the internet in 2017 and in January 2018, a desktop application called FakeApp was launched. Similarly, voice mimicking software such as Lyrebird or Baidu's Deep Voice can "clone" anyone's voice. The Chinese tech giant application only needs 3.7 s of audio of a voice to reproduce it [25]. The combination of voice and image forgery will make any piece of media on the internet suspicious. Such applications have democratized the ability to create perfect visual and audio manipulations [26]. This is often referred to as the "end of truth" or the end of "seeing is believing", which Henry Kissinger has identified as leading to the "end of the Enlightenment era" [27]. Building such algorithms without security in mind, and without thinking about the possible repercussions on society carries enormous risk.

15.5 Social Control and Discrimination

As mentioned earlier, algorithms have surpassed humans at image recognition, which means that AI is much better at identifying visual patterns, including for facial recognition. Some governments have seen the benefits of such technologies and use it to increase the surveillance of their citizens. China has gone the furthest in this field. AI-enabled technologies have allowed Beijing to create an advanced surveillance system by awarding Chinese citizens a social score based on their online and offline behaviour. As Rickli stated previously, "the Chinese government has implemented a surveillance system based on the gamification of obedience through big data and artificial intelligence. It relies on punitive and reward measures that influence the way its citizen should behave (quoted in [28])".

Beyond this, the Chinese government is also using facial recognition algorithms to identify one specific ethnic group, the Uighurs, for law enforcement purposes. The Uighurs are a minority of 11 million, mostly located in the western region of Xinjiang. China is mainly populated by the Han ethnic group. The Chinese police has used "facial recognition technologies to target Uighurs in wealthy eastern cities like Hangzhou and Wenzhou and across the coastal province of Fujian" and it is spreading to more than 16 different provinces and regions across China [29]. In one city, law enforcement authorities ran such a system more than 500,000 times within the course of a month in 2019 to screen whether residents were Uighurs. The purpose of this technology is to monitor and track this ethnic group, which the Chinese government accuses of ethnic violence and terrorist attacks.

Ethnic profiling is a dangerous development in facial recognition technologies and AI more generally that is very appealing to authoritarian regimes. Chinese AI surveillance technologies are now also being exported to other states such as Zimbabwe, Singapore, Malaysia or Mongolia [29].

15.6   Military Applications of AI

The military domain is not immune to developments in AI. With artificial intelligence, the new tactic of swarming will become possible in the physical domain. Swarming relies on overwhelming and saturating the adversary's defence system by synchronizing a series of simultaneous and concentrated attacks [30]. In October 2016, the US Department of Defense conducted an experiment that saw 103 Perdrix micro drones autonomously deal with four different objectives. Meanwhile, the world record for swarming drones was broken by a Chinese company, EHang, in May 2018 with an AI-assisted swarm of 1374 drones flying over the City wall of Xi'An and then by the US company Intel in July 2018 with 2018 drones [31, 32]. Swarming tactics are potentially disruptive because they combine firepower, mass and speed.

These factors combined with the specific capabilities of artificial narrow intelligence systems means that defence is rapidly becoming costlier and less effective than offence, shifting the dynamics of security towards pre-emption [33].

The development of lethal autonomous weapons systems (LAWS) in particular will likely have a destabilizing impact on strategic stability in the future [34]. Since WWII, strategic stability has been guaranteed by the supremacy of the defensive, especially due to the sheer destructive power of the second-strike retaliatory capabilities of nuclear weapons. If the applications of swarming tactics make second-strike retaliatory capabilities an illusion because of the offensive advantage provided by swarming, it will follow that deterrence will be replaced by pre-emption. These changes in strategy are very likely to create an unstable international configuration that encourages escalation and arms races [35].

So far, international law prohibits the use of military force except in cases of self-defence and if the UN Security Council allows it under Chap. 7 of the UN Charter. If the offensive has the advantage, the only way to protect yourself is by attacking first. Pre-emption is therefore in direct contradiction to the spirit of the UN Charter and its application is a violation of Art 2(4) of the Charter. As Rickli argues, this new international system that stems from the militarisation of AI will be much more unstable and prone to conflicts and will make pre-emption the strategy of choice to deal with adversaries [36].

Moreover, the growing use of autonomy in weapon systems allows the potential development of weapons that will be fully autonomous. These weapons will be able to move independently through their environment to arbitrary locations, select and fire upon targets in their environment and create and/or modify its goals, incorporating observation of its environment and communication with other agents [37]. Such weapons will accelerate a trend in the development of warfare in the twenty-first century, which entails that state and non-state actors increasingly rely on both human and technological surrogates to fight on their behalf [38]. Such developments favour international instability because it reduces the threshold to use force as well as a drastic reduction in the accountability of the use of force.

15.7 Security Implications of Democratization of Access

A key characteristic of emerging technologies is the rapid decrease in the cost of access [39]. In the case of AI, the drop in the cost of the technology is due to the growth of the processing power of CPUs and the creation of larger data sets. Furthermore, the digital nature of AI systems—and the fact that AI algorithms are often public or even open-source—allows them to be distributed and scaled rapidly [40].

As a result of these cost shifts, lower barriers to entry incentivize new actors to use this technology. From a security perspective, the automation of tasks mean that individuals will potentially become more dangerous as they may have access to technologies with disruptive impacts. As greater numbers of actors invest in AI-driven tactics, higher rates of experimentation and innovation will result in the emergence and proliferation of new threats and tactics [41].

The falling costs and the accessibility to AI particularly empowers individuals, small groups, criminal enterprises and other non-state actors [42]. This is very visible in the cyber domain, where the acquisition of new cyber capabilities is cheap and the marginal cost of additional production—adding a target—is close to zero [43]. Equally, in the physical domain, AI-enabled commercial products can easily be repurposed for surveillance purposes or to attack targets [40]. Although not AI, ISIS mounted high-definition cameras under drones to improve intelligence and acquire situational awareness during their combat operations. They also used drones to drop 40 mm grenades on Iraqi positions, allegedly killing up to 30 Iraqi soldiers per week during the battle of Mosul in 2017 [44]. This demonstrates how agile terrorist organizations are in using commercial technologies to support their goals. AI will probably not be an exception to the rule in that once algorithms have been developed they are either easily accessible once they are released into databases (e.g. Tensorflow) or can be deducted from adversarial black-box attacks. The next section looks at the security implications of neurotechnology.

15.8   The Security Implications of Neurotechnology

Neurotechnology can be defined as "devices and procedures that are used to access, monitor, investigate, assess, manipulate and emulate the structure and function of neural systems" [45, 46]. While AI systems emulate or simulate functional aspects of the (human) brain, neurotechnologies are designed to record, monitor, functionally understand and modulate processes in the (human) brain. Neurotechnologies stricto-sensu include non-invasive medical imaging technologies such as magnetic resonance imaging (MRI) and near-infrared spectroscopy (NIRS), electrode-based electrophysiological monitoring (EEG), non-invasive neuromodulation techniques such as transcranial magnetic stimulation (TMS) or transcranial electric stimulation (tES), sensory neuroprosthetics such as visual or auditory prostheses as well as invasive neurostimulation techniques involving implant neurosurgery such as deep brain stimulation (DBS). Broader definitions of neurotechnology also encompass computational simulations of neural functions and neuromorphic engineering.

Neurotechnology originated in the clinical domain as an array of tools and techniques aimed at monitoring, modulating, restoring or enhancing neural structures or functions. Furthermore, neurotechnology plays a critical role in research and is a major enabler of discovery and translational neuroscience. Advances in neurotechnology are necessary requirements for achieving the grand challenges of contemporary neuroscience, namely: (a) reliably measuring neuronal activity, (b) mapping neuronal activity onto a reliable and highly detailed anatomical and functional atlas of the brain and (c) making sense of the brain by mining large volumes of brain data through reliable and high-velocity analytic techniques [47]. Meeting these three scientific challenges, in turn, is essential to the development of preventative, diagnostic, therapeutic or assistive solutions that might reduce the burden of neurological disorders and improve the lives of millions of patients.

In recent years, advances in neuroengineering and pervasive computing, combined with increased extra-clinical interest in the potential of neurotechnology, have propelled neurotechnologies from the exclusive clinical and biomedical domains onto a broad variety of commercial [48], educational [49] and military applications. Consumer-grade neurotechnologies include several non-invasive neurodevices such as neuromodulatory devices based on transcranial direct current stimulation (tDCS) or transcranial magnetic stimulation (TMS), brain–computer interfaces (BCIs) for self-neuromonitoring and device control, and an associated ecosystem of both proprietary and open-source software (including mobile applications).

The proliferation of neurotechnology outside clinics and research domains raises security implications. The reason is threefold. First, the domain of consumer neuro-technologies is, to date, largely unregulated. While clinical neurotechnologies are subject to medical device regulation and stricter privacy rules for the processing of health data, consumer neurotechnologies are currently being developed in an unde-fined legal territory, and existing regulatory oversight has been deemed "insuffi-cient" by experts [48, 50]. The absence of adequate oversight mechanisms and unambiguous regulation increases the chances that security breaches might emerge [51], some of which reportedly already have [52]. Furthermore, unlike clinical and research applications, consumer neurotechnologies are not typically used in a medi-cally supervised environment and are not subject to continuous safety monitoring by researchers. This increases the chances that the technology might be misused either by the users themselves or by third parties. Finally, the proliferation of unsu-pervised neurotechnology applications causes a proliferation of actors involved in the handling of neurodevices and derived brain data. Today, the categories of actors involved in the development and use of neurotechnology do not exclusively com-prise neuroscientists, neuroengineers and neurological patients. Consumer neuro-technology applications have opened the gates of neurotechnology use to the general population, including healthy individuals. Furthermore, following sociotechnical trends such as do-it-yourself (DIY) neurotechnology and biohacking, neurotech-nologies are increasingly being developed and experimented with by non-professional scientists. These trends are causing both a proliferation of actors and a fragmentation of oversight measures, with a consequent increase in security risks.

To comprehensively map the dual-use landscape of neurotechnology, it should be noted that the extra-clinical proliferation of neurotechnology is not limited to the civilian domain, but also extends to the military sector. In the last decade, several neurotechnologies have gained ground as experimental applications among govern-mental national security agencies such as the Defense Advanced Research Projects Agency (DARPA), a research agency of the United States Department of Defense. Military uses of neurotechnology include experimental applications for brain-based intercept-proof communication, remote device control (e.g. brain-controlled unmanned aerial vehicles), warfighter enhancement and post-traumatic treatment of veterans. This process of permeation of neurotechnology in the state military sector has been termed "weaponization of neuroscience" [53], even though authors have argued that neurotechnology has been "a toll of war from the start" [54]. For exam-ple, Howell has observed that the origin of clinical neurology is intertwined with the American civil war and that the birth of modern neuroscience was highly dependent on research conducted with military research institutions such as the Walter Reed Army Institute of Research (ivi). Finally, misuse of neurotechnology by malign non-state actors has also been indicated as a primary source of risk for international security [52, 55].

15.9 Data Bias, Agency and Accountability

As the functioning of neurotechnology, especially BCI, is highly dependent on data, data quality and data protection measures are paramount to ensure safety and security. Biases in datasets, poor data quality and corrupted data can all negatively affect the functioning of neurotechnologies and lead to suboptimal or even harmful outcomes. Furthermore, experts have argued that algorithmic biases, such as those affecting datasets used to feed AI applications, could become embedded in neural devices [56]. The reason for this stems from the fact that most neurotechnologies rely on machine learning and other AI techniques to decode brain signals and translate them into utilizable output. Consequently, biases contained in the datasets used to train those algorithms are likely to be transferred or even amplified during the process. This risk is exacerbated in the context of neurotechnologies used by vulnerable user groups such as children, patients with neurological disorders or socially marginalized individuals.

The increasing use of machine learning and, more generally, of artificial intelligence to optimize BCI functions also has implications for the notion of action and responsibility. For example, Klaming and Haselager [57] have hypothesized that when BCI control is partly dependent on intelligent algorithmic components, it may become difficult to discern whether the resulting behavioural output was actually executed by the user. This difficulty introduces a principle of indetermination into the cognitive process that starts from the conception of an action (or intention) to its execution, with consequent uncertainty in attributing responsibility to the author of such action. This principle of indetermination could call into question the notion of individual responsibility, with obvious consequences of a criminal and insurance nature. More broadly, it could also jeopardize the entire concept of legal liability because liability is predicated upon the state of a legal person of being legally responsible. If the intelligent components embedded in the BCI override the human user's volition or simply make any discrete attribution of responsibility indeterminable, this would represent a fundamental transformation of both the civil and criminal law systems as they both rely on the establishment of liability to make actors responsible or answerable in law. Moreover, the principle of indetermination could generate a sense of estrangement in the user, whose ethical relevance is all the greater if he/she is a vulnerable individual such as a neurological patient. In addition, there is a possibility that the centrality of these intelligent components in the functioning of the BCI may affect the user's subjective experience, and thus their personal identity [58]. This hypothesis has recently obtained a preliminary empirical confirmation in a qualitative study about the personal experience of DBS patients [59].

15.10 Manipulations

Unlike disembodied AIs, manipulation risks associated with neurotechnology involve the modification of underlying neurobiological functioning for the obtainment of emotional, cognitive or behavioural aims. An example is research on neurotechnology for selective memory manipulation. Nabavi and colleagues used an optogenetics technique to erase and subsequently restore selected memories by applying a stimulus via optical laser that selectively strengthens or weakens synaptic connections [60]. As noted by Ienca and Andorno [61], the future sophistication and misuse of these techniques by malevolent actors may generate unprecedented opportunities for mental manipulation and brain-washing [61]. In particular, it has been observed that neurostimulation may have an impact on the psychological continuity of the person, i.e. the crucial requirement of personal identity consisting in experiencing oneself as persisting through time as the same person [57]. Consequently, by using neurostimulation it is possible, in principle, to manipulate the psychology of a person in manners that might affect that person's identity. It has been reported, for example, that invasive BCIs, such as DBS, may lead to behavioural changes such as increased impulsivity and aggressiveness [62], different taste in music [63] or changes in sexual behaviour [64]. Such induced behavioural changes might be of potential interest for state and non-state actors.

More subtle forms of manipulation based on non-invasive neurotechnology have also been discussed in the literature. An example is unconscious neural advertising via neuromarketing. Neuromarketing allows the use of techniques such as embedding subliminal stimuli with the purpose of eliciting responses (e.g. preferring item A instead of B) that people cannot consciously register. This has raised criticism among consumer advocate organizations, such as the Center for Digital Democracy, which have warned against neuromarketing's potentially manipulative application. Jeff Chester, the executive director of the organization, has claimed that "though there has not historically been regulation on adult advertising due to adults having defense mechanisms to discern what is true and untrue", it should be regulated "if the advertising is now purposely designed to bypass those rational defenses" [65].

15.11 Social Control and Discrimination

Neuromonitoring technology is vulnerable to the risk of being co-opted for surveillance and social control. The South China Morning Post has reported, for example, that in China state-backed neuroheadsets for EEG-based neuromonitoring are being deployed to detect changes in emotional states in three categories of individuals: public employees on the production line, the military and conductors of high-speed trains on the Beijing-Shanghai rail line [66]. Compelled use of neuromonitoring technology has raised concerns in terms of cognitive liberty and mental privacy. Authors have argued that every individual should be free to decide whether to use a certain neurotechnology application or refuse to do so, hence that coercive use should be prohibited [61]. Furthermore, the informational richness of brain data and their localization under the threshold of conscious control make it difficult for neurotechnology users to consciously segregate the information they want to seclude from what they want to share. Therefore, there is a risk that neuromonitoring activities might cause privacy breaches into a person's psychological life, hence resulting in violations of mental privacy. Mental privacy breaches can lead to discrimination in a twofold manner: either as a result of bias contained in the datasets or as the purposive extraction from brain recordings of predictive information about health status and behaviour. For example, neural signatures of Alzheimer's disease or risk-taking behaviour can be used to discriminate individuals in manners that range from job termination to increased insurance premiums.

15.12   Military Applications of Neurotechnology

According to Tennison and Moreno, military applications of neurotechnology fall into three main categories: brain-computer interfaces (BCIs), neurotechnologies for warfighter enhancement, and neurotechnological systems for deception detection and interrogation [67]. The first category encompasses systems that establish a direct connection channel between the human brain and an external computer device, bypassing the peripheral nervous and muscular system. Military uses of BCIs include the acquisition of neural information gathered from warfighters' brains to adaptively modify their equipment and the development of threat warning systems that convert subconscious, neurological responses to danger into consciously available information [68]. Some authors refer to "disruptive BCIs" when they are planned to be used in an offensive manner, especially in a military setting such as the degradation and/or reading of enemy cognitive, sensory, motor neural activity [69]. These BCIs could be used, in the future, for torture or interrogation purposes, raising particular ethical questions.

Warfighter enhancement applications include various forms of transcranial electric stimulation technology such as transcranial direct current stimulation (tDCS) for selective cognitive enhancement in targeted brain areas. Finally, the deception detection domain encompasses devices such as the so-called "brain-fingerprints" capable of accessing concealed information in response to a stimulus. While these applications, especially those based on functional magnetic resonance (fMRI) and electroencephalography (EEG), hold great potential for medical diagnostics, they can be used as surveillance and interrogation tools for national security purposes. Unlike more rudimental interrogation technologies such as polygraph-based lie detection (based on the recording of extra-cranial physiological indices such as pulse and skin conductivity), brain-based lie detection technologies associate the truth-values of an uttered sentence or a mental state with specific patterns of brain activity.

The rise of network-centric warfare, a networked form of warfare relying on digital technologies, has increased the prevalence of hacking as a real threat to the capacity of armed forces to conduct operations. This concern can be extended to BCIs in ways which can be even more unsettling as we are speaking of hacking the cognitive, emotional and life support functions of humans. This risk opens the prospect of "malicious brain-hacking", namely the "possibility of co-opting brain-computer interfaces and other neural engineering devices with the purpose of accessing or manipulating neural information from the brain of users" [52]. The ability to penetrate human brains through BCI will in fact add a new dimension to physical and cyber security and warfare in the future. This could, in the distant future, potentially lead to weapons that could "capture minds" (for example, via selective memory manipulation, coercive neurostimulation or brain-to-brain control) with consequent implications not only for biosecurity [70], but also for human rights [61]. Artificial intelligence approaches such as deep learning have already been successfully used for neural control purposes in animal models involving monkeys [71].

15.13 Security Implications of Democratization of Access

As a consequence of decreasing hardware costs, improvements in sensorics and the increasing feasibility of developing portable EEG, functional near-infrared spectroscopy (fNIRS), transcranial electrical stimulation (tES) and transcranial magnetic stimulation (TMS) based neurotechnologies, the neurotechnology spectrum is not restricted to clinical and research applications, but includes a wide variety of direct-to-consumer systems [48]. This consumer neurotechnology trend is determining a proliferation and democratization of actors involved in the utilization of neurotechnologies. For instance, commercially available EEG-based consumer neurotechnologies start at about €120, hence making them affordable for many individuals globally [72]. Another sociotechnical trend known as do-it-yourself (DIY) neurotechnology has empowered non-professional individuals (often self-proclaimed biohackers) to self-assemble neurotechnology devices for personal use, most frequently via transcranial electrical stimulation for self-improvement purposes. Furthermore, as DTC neurotechnologies are typically utilized in absence of medical or other professional supervision, this proliferation also implies a reduced ability of authorities to monitor who is using neurotechnologies, how they are being used and for which purposes. Democratizing cognitive technology, neurotechnology in particular, is a laudable and to-be-pursued ethical goal [2] because it favours fair access and the just distribution of the societal benefits of this technology. Furthermore, it minimizes the risk that advantaged individuals, organized groups or states could achieve disproportionate control over the technology and use it for personal gain, surveillance or social control purposes at the expense of the majority of the population. At the same time, however, the proliferation of actors and the increased opacity of neurotechnology uses increase the statistical probability that these technologies might be used by malevolent actors for non-benign purposes. In light of these trends, authors have highlighted the urgent need for more agile, adaptive and systemic oversight mechanisms, neurosecurity standards, global governance frameworks and ethically aligned design via responsible innovation [2, 48, 50, 55, 70, 73].

15.14 Conclusion

This article has illustrated that the two families of cognitive technology, namely artificial intelligence and neurotechnology, are not only converging in terms of development and applicability, but also raising parallel security implications. In fact, both technologies hold great transformative potential due to their ability to read, modify and amplify human cognition in a variety of domains and in response to a variety of cognitive and analytical tasks. Furthermore, both neurotechnology and artificial intelligence are rapidly proliferating outside of traditional supervised settings (e.g. clinics and academic research) onto multiple and unsupervised domains, a phenomenon that can be labelled "horizontal proliferation". Among these domains, their co-optation into the military sector and subsequent weaponization are of particular concern from an international security perspective. Similarly, as it is the case with artificial intelligence, proliferation also happens vertically, from state to non-state actors and individuals and vice versa. This is because of the dual-use nature of these technologies. Thus, it is extremely difficult to monitor and control the way they are used and, more importantly, misused. Indeed, with the ease of proliferation, one cannot exclude that these technologies will be used for malevolent purposes. This can already be observed with AI and deepfakes used to purposely modify satellite pictures, for instance [74].

In light of their disruptive potential and rapid proliferation, both neurotechnology and artificial intelligence urge global governance responses that deal with their accessibility, their proliferation, their dual-use nature including how easily these technologies can be repurposed and obviously, the ethics and values that should accompany the development and use of these technologies. These responses should be inclusive and comprise all the different stakeholders (governments, private sector, scientific community, civil society and tech companies) and be very versatile in that these technologies and applications evolve rapidly.

Acknowledgments Jean-Marc Rickli would like to thank Federico Mantellassi and Alexander Jahns for the background research conducted. Marcello Ienca would like to thank Fabrice Jotterand and Ralf Jox for their insightful comments to the research presented in this chapter.

Author contributions: JMR & MI conceived of the study and wrote the chapter. The two authors contributed equally.

References

1. Dascal M, Dror IE. The impact of cognitive technologies: towards a pragmatic approach. Pragmat Cogn. 2005;13(3):451–7.
2. Ienca M. Democratizing cognitive technology: a proactive approach. Ethics Inf Technol. 2019;21(4):267–80.
3. Ingham L. Stephen Hawking: the rise of powerful AI will be either the best or the worst thing ever to happen to humanity. A factor. 2018. https://www.factor-tech.com/feature/stephen-hawking-the-rise-of-powerful-ai-will-be-either-the-best-or-the-worst-thing-ever-to-happen-to-humanity/.
4. Moore BE. The brain computer interface future: time for a strategy. A research report submitted to the faculty. Air War College: Air War College Air University Maxwell AFB United States; 2013. https://apps.dtic.mil/dtic/tr/fulltext/u2/1018886.pdf.
5. Elements of AI. How should we define AI. 2019. https://course.elementsofai.com/1/1.
6. Independent High Level Expert Group on Artificial Intelligence. A definition of AI: main capabilities and disciplines: Brussels, European Commission; 2018.
7. SAS. Artificial intelligence, what it is and why it matters. 2019. https://www.sas.com/en_us/insights/analytics/what-is-artificial-intelligence.html.

8. Jajal TD. Distinguishing between narrow AI, general AI and super AI. Medium; 2018.
9. Grace K, Salvatier J, Dafoe A, Zhang B, Evans O. When will AI exceed human performance? Evidence from AI experts. J Artif Intell Res. 2018;62:729–54.
10. Bostrom N. Superintelligence. Paths, dangers, strategies. Oxford: Oxford University Press; 2014.
11. Deepmind. AlphaStar: mastering the real-time strategy game StarCraft II. 2019. https://deepmind.com/blog/alphastar-mastering-the-real-time-strategy-game-starcraft-ii/.
12. Service RF. Google's deepmind aces protein folding. Science. 2018. https://www.sciencemag.org/news/2018/12/google-s-deepmind-aces-protein-folding.
13. Metz C. How Google's AI viewed the move no human could understand. Wired. 2016.
14. Brown N, Sandholm T, editors. Libratus: the superhuman AI for no-limit poker. In: Twenty-sixth international joint conference on artificial intelligence (IJCAI-2017); 2017.
15. Open AI. AI and compute. San Francisco: OpenAI; 2018.
16. Open AI. Better language models and their implications. San Francisco: OpenAI; 2019.
17. Agence France Press. Computer learns to detect skin cancer more accurately than doctors. The Guardian. 2018.
18. Yarnardag P. Normann: world's first psychopath AI. Cambridge: MIT; 2018. http://norman-ai.mit.edu.
19. Angwin J, Larson J, Mattu S, Kirchner L. Machine bias. Pro Publica. 2016. https://www.propublica.org/article/machine-bias-risk-assessments-in-criminal-sentencing.
20. Resnick B. Yes, artificial intelligence can be racist. VOX. 2019.
21. Kahn J. Major tech firms come out against police use of algorithms. Bloomberg. 2019. https://www.bloomberg.com/news/articles/2019-04-26/major-tech-firms-come-out-against-police-use-of-ai-algorithms.
22. Sentient. Understanding the "blackbox" of artificial intelligence. San Francisco: Sentient Technologies Holdings Limited; 2018. https://www.sentient.ai/blog/understanding-black-box-artificial-intelligence/.
23. Henschen D. How ML and AI will transform business intelligence analytics. ZDNet. 2018. https://www.zdnet.com/article/how-machine-learning-and-artificial-intelligence-will-transform-business-intelligence-and-analytics/.
24. Hern A. Cambridge analytica: how did it turn clicks into votes. The Guardian. 2018. https://www.theguardian.com/news/2018/may/06/cambridge-analytica-how-turn-clicks-into-votes-christopher-wylie.
25. Cole S. Deep voice software can clone anyone's voice with just 3.7 seconds of audio. Motherboard. 2018. https://motherboard.vice.com/en_us/article/3k7mgn/baidu-deep-voice-software-can-clone-anyones-voice-with-just-37-seconds-of-audio.
26. Cauduro A. Live deep fakes—you can now change your face to someone else's in real time video applications. Medium. 2018. https://medium.com/huia/live-deep-fakes-you-can-now-change-your-face-to-someone-elses-in-real-time-video-applications-a4727e06612f.
27. Kissinger H. How the enlightenment ends. Atlantica. 2018. https://www.theatlantic.com/magazine/toc/2018/06/.
28. Joplin T. Long form: China's global surveillance-industrial complex. Albawaba News. 2018. https://www.albawaba.com/news/long-form-china's-global-surveillance-industrial-complex-1141152.
29. Mozur P. One month, 500,000 face scans: how China is using A.I. to profile a minority. Ney York Times. 2019. https://www.nytimes.com/2019/04/14/technology/china-surveillance-artificialintelligence-racial-profiling.html.
30. Scharre P. Robotics on the battlefield part II. Washington, DC: Center for a New American Security; 2014.
31. EHang. EHang Egret's 1374 drones dancing over the city wall of Xi'an, achieving a Guinness World Title. 2018. https://www.ehang.com/news/365.html.
32. Weaver D, Black E. Behind the scenes as Intel sets the world record for flying over 2000 drones at once. CNBC. 2018. https://www.cnbc.com/2018/07/17/intel-breaks-world-record-2018-drones.html.

33. Rickli J-M. The destabilizing prospects of artificial intelligence for nuclear strategy, deterrence and stability. In: Boulanin V, editor. The impact of artificial intelligence on strategic stability and nuclear risk: European perspectives. I. Stockholm: Stockholm International Peace Research Institute; 2019. p. 91–8. https://www.sipri.org/sites/default/files/2019-05/sipri1905-ai-strategic-stability-nuclear-risk.pdf.
34. Altmann J, Sauer F. Autonomous weapon systems and strategic stability. Survival. 2017;59(5):117–42.
35. Rickli J-M. The impact of autonomous weapons systems on international security and strategic stability. In: Ladetto Q, editor. Defence future technologies: what we see on the horizon. Thun: Armasuisse; 2017. p. 61–4. https://deftech.ch/What-We-See-On-The-Horizon/armasuisseW%2BT_Defence-Future-Technologies-What-We-See-On-The-Horizon-2017_HD.pdf.
36. Rickli J-M. The impact of autonomy and artificial intelligence on strategic stability. UN Special. 2018. p. 32–3. https://www.unspecial.org/2018/07/the-impact-of-autonomy-and-artificial-intelligence-on-strategic-stability/.
37. Roff H, Moyes R. Autonomy, robotics and collective systems. Tempe: Global Security Initiative, Arizona State University; 2016. https://globalsecurity.asu.edu/robotics-autonomy.
38. Krieg A, Rickli J-M. Surrogate warfare: the transformation of war in the twenty-first century. Georgetown: Georgetown University Press; 2019.
39. Rickli J-M. Education key to managing risk of emerging technology. European CEO. 2019. https://www.europeanceo.com/industry-outlook/education-key-to-managing-the-threats-posed-by-new-technology/.
40. Davis N, Rickli J-M. Submission to The Australian Council of Learned Academies and the Commonwealth Science Council on the opportunities and challenges presented by deployment of artificial intelligence. ACLO, Melbourne 2018.
41. Brundage M, Avin S, Clark J, Toner H, Eckersley P, Garfinkel B, et al. The malicious use of artificial intelligence: forecasting, prevention, and mitigation. arXiv preprint arXiv:180207228. 2018. https://arxiv.org/abs/1802.07228.
42. Rickli J-M. The economic, security and military implications of artificial intelligence for the Arab Gulf Countries. Emirates Diplomatic Academy Policy Paper. 2018. https://www.gcsp.ch/News-Knowledge/Global-insight/The-Economic-Security-and-Military-Implications-of-Artificial-Intelligence-for-the-Arab-Gulf-Countries.
43. Allen G, Chan T. Artificial intelligence and national security. Cambridge: Belfer Center for Science and International Affairs; 2017. https://www.belfercenter.org/sites/default/files/files/publication/AI%20NatSec%20-%20final.pdf.
44. Chovil P. Air superiority under 2000 feet: lessons from waging drone warfare against ISIL. War on the Rocks. 2018. https://warontherocks.com/2018/05/air-superiority-under-2000-feet-lessons-from-waging-drone-warfare-against-isil/.
45. Garden H, Bowman DM, Haesler S, Winickoff DE. Neurotechnology and society: strengthening responsible innovation in brain science. Neuron. 2016;92(3):642–6.
46. Giordano J. Neurotechnology: premises, potential, and problems. Boca Raton: CRC Press; 2012.
47. Abbott A. Neuroscience: solving the brain. Nature. 2013;499(7458):272.
48. Ienca M, Haselager P, Emanuel EJ. Brain leaks and consumer neurotechnology. Nat Biotechnol. 2018;36:805.
49. Behneman A, Berka C, Stevens R, Vila B, Tan V, Galloway T, et al. Neurotechnology to accelerate learning: during marksmanship training. IEEE Pulse. 2012;3(1):60–3.
50. Wexler A, Reiner PB. Oversight of direct to consumer neurotechnologies. Science. 2019;363(6424):234–5.
51. Dupont B. Cybersecurity futures: how can we regulate emergent risks? Technol Innov Manag Rev. 2013;3(7):6–11.
52. Ienca M, Haselager P. Hacking the brain: brain-computer interfacing technology and the ethics of neurosecurity. Ethics Inf Technol. 2016;18(2):117–29.

53. Walther G. Weaponization of neuroscience. In: Clausen J, Levy N, editors. Handbook of neuroethics. Dordrecht: Springer; 2015. p. 1767–71.
54. Howell A. Neuroscience and war: human enhancement, soldier rehabilitation, and the ethical limits of dual-use frameworks. Millennium. 2017;45(2):133–50.
55. Ienca M, Vayena E. Dual use in the 21st century: emerging risks and global governance. Swiss Med Wkly. 2018;148:w14688.
56. Yuste R, Goering S, Agüera y Arcas B, Bi G, Carmena JM, Carter A, et al. Four ethical priorities for neurotechnologies and AI. Nature. 2017;551(7679):159–63.
57. Klaming L, Haselager P. Did my brain implant make me do it? Questions raised by DBS regarding psychological continuity, responsibility for action and mental competence. Neuroethics. 2013;6(3):527–39.
58. Ferretti A, Ienca M. Enhanced cognition, enhanced self? On neuroenhancement and subjectivity. J Cogn Enhancement. 2018;2(4):348–55.
59. Gilbert F. Deep brain stimulation: inducing self-estrangement. Neuroethics. 2018;11(2):157–65.
60. Nabavi S, Fox R, Proulx CD, Lin JY, Tsien RY, Malinow R. Engineering a memory with LTD and LTP. Nature. 2014;511(7509):348–52.
61. Ienca M, Andorno R. Towards new human rights in the age of neuroscience and neurotechnology. Life Sci Soc Policy. 2017;13(1):1–27.
62. Frank MJ, Samanta J, Moustafa AA, Sherman SJ. Hold your horses: impulsivity, deep brain stimulation, and medication in parkinsonism. Science. 2007;318(5854):1309–12.
63. Mantione M, Figee M, Denys D. A case of musical preference for Johnny Cash following deep brain stimulation of the nucleus accumbens. Front Behav Neurosci. 2014;8:152.
64. Houeto JL, Mesnage V, Mallet L, Pillon B, Gargiulo M, du Moncel ST, et al. Behavioural disorders, Parkinson's disease and subthalamic stimulation. J Neurol Neurosurg Psychiatry. 2002;72(6):701–7.
65. Singer N. Making ads that whisper to the brain. New York Times. 2010.
66. Chen S. Forget the Facebook leak: China is mining data directly from workers' brains on an industrial scale. South China Morning Post. 2018.
67. Tennison MN, Moreno JD. Neuroscience, ethics, and national security: the state of the art. PLoS Biol. 2012;10(3):e1001289.
68. Miranda RA, Casebeer WD, Hein AM, Judy JW, Krotkov EP, Laabs TL, et al. DARPA-funded efforts in the development of novel brain–computer interface technologies. J Neurosci Methods. 2015;244:52–67.
69. Munyon CN. Neuroethics of non-primary brain computer interface: focus on potential military applications. Front Neurosci. 2018;12:696.
70. Ienca M, Jotterand F, Elger BS. From healthcare to warfare and reverse: how should we regulate dual-use neurotechnology? Neuron. 2018;97(2):269–74.
71. Bashivan P, Kar K, DiCarlo JJ. Neural population control via deep image synthesis. Science. 2019;364(6439):eaav9436.
72. Wexler A. The social context of "do-it-yourself" brain stimulation: neurohackers, biohackers, and lifehackers. Front Hum Neurosci. 2017;11:224.
73. Goering S, Yuste R. On the necessity of ethical guidelines for novel neurotechnologies. Cell. 2016;167(4):882–5.
74. Tucker P. The newest AI-enabled weapon: deep-faking photos of the earth. Defense One. 2019. https://www.defenseone.com/technology/2019/03/next-phase-ai-deep-faking-whole-world-and-china-ahead/155944/.

Connecting Brain and Machine: The Mind Is the Next Frontier

16

Mathias Vukelić

Contents

Abstract

Artificial intelligence coupled with digitally connected technologies are becoming more self-evident. These developments indicate an increasing symbiosis between human and machine, referring to a new phase of interaction—symbiotic intelligence. In this vein, the human-centred development of technologies is becoming more and more important. The detection of user's mental states, such as cognitive processes, emotional or affective reactions, offers great potential for the development of intelligent and interactive machines. Neurophysiological signals provide the basis to estimate many facets of subtle mental user states, like attention, affect, cognitive workload and many more. This has led to extensive progress in brain-based interactions—Brain-Computer Interfaces (BCIs). While most BCI research aims at designing assistive, supportive or restorative systems for severely disabled persons, the current discussion focuses on neuroadaptive control paradigms using BCIs as a strategy to make technologies more human-centred and also usable for non-medical applications. The primary goal of our

M. Vukelić (✉)
Fraunhofer Institute for Industrial Engineering IAO, Stuttgart, Germany
e-mail: mathias.vukelic@iao.fraunhofer.de

© Springer Nature Switzerland AG 2021
O. Friedrich et al. (eds.), *Clinical Neurotechnology meets Artificial Intelligence*,
Advances in Neuroethics, https://doi.org/10.1007/978-3-030-64590-8_16

neuroadaptive technology research agenda is to consistently align the increasing intelligence and autonomy of machines with the needs and abilities of the human—a human-centred neuroadaptive technology research roadmap. Due to its far-reaching social implications, our research and developments do not only face technological but also social challenges. If neuroadaptive technologies are applied in non-medical areas, they must be consistently oriented to the needs and ethical values of the users and society.

16.1   The Rise of Artificial Intelligence: Technologies for the Interaction Between Human and Machine

Digitally connected systems, techniques and methods of artificial intelligence (AI) and machine learning (ML) are changing our world. Humans develop such technologies in order to satisfy intentions, thereby anticipating our needs and thus making life easier. Computers trade our shares, we have cars that park themselves, and flying is almost completely automated. Virtually every area has benefited from the tremendous progress in digitization and AI, from military to the medical field and manufacturing. These progresses not only gradually transform the way we are interacting with technological products and services, but also differently influences our sensorimotor and cognitive capacities and skills. Osiurak et al. [1] summarizes these gradual technological developments over time into three levels that describe human-technology interaction with physical (affordance design), sophisticated (automation and interface design) and symbiotic (embodied and cognitive design) technologies. In the future, we will experience a massive influence on people's everyday lives and working environment: the interaction with digital products and connected machines, technologies and services is becoming more self-evident and a core competence of the future requiring new modes of interaction and cognitive abilities. Future trends, like voice, gesture or thought operated technologies indicate an increasing symbiosis of human and technology, referring to a new phase of interaction called *symbiotic intelligence* [1]. The authors claim that the sophisticated technology of the future will ultimately become more and more unconscious to humans in order to maybe become one with them—the goal is the intuitive handling of technology in order to minimize the interaction effort with technical products.

Historically, the development and use of tools is strongly related to human evolution and intelligence. With the rise of AI and digitally connected products, we have access to an enormous variety of data and information. This enables us to develop interfaces that support us in how we think, what we know, how we decide and act. This transformation can be summarized under the concept of cognitive enhancement or cognitive augmentation [2]. It describes a very broad spectrum of techniques and approaches, such as performance-enhancing drugs, medical implants and prostheses and human-computer interfaces, which lead to improved abilities and may probably transcend our existing cognitive boundaries. Nevertheless, the increasing integration of technology in our everyday life and working environments entails new challenges and potential for conflicts. Often, humans with their individual preferences, skills and needs find themselves overlooked in the development of future technology. The resulting solutions, while technologically advanced, may nevertheless offer limited gains in terms of the productivity, creativity, and health of the users in question.

16.2   Embodied and Situated Minds: How We Use, Act and Think with Technology

If smart and adaptive technologies that support or even expand our cognitive abilities are the future, then it is essential to consider an optimal design so that such technologies are geared to the user's needs and contribute to a human-centred, efficient and accepted technology.

The human-centred development of technologies and interfaces for the interaction between human and machines is becoming more and more important. In order to achieve increased productivity with a concurrent contribution to the subjective well-being of employees, digital equipment needs to be seamlessly and intuitively integrated in everyday working life [3]. Intelligent systems should support the user rather than hamper the interaction due to its inherent complexity. Instead of creating frustration, the system should motivate the user by providing a positive user experience during the interaction [4]. Positive user experiences in daily human-technology interactions are extremely important for both the individual person and the organization: from the human factors point of view, positive user experiences contribute to the subjective perception of competence, have a positive effect on mental health and consequently lead to motivated action, increased productivity and job satisfaction, which are important factors for the enterprise [5–7].

Research from the cognitive neurosciences on embodied intelligence shows that intelligence cannot be assigned purely to brain functions without regarding the human in its situated surroundings. Thus, intelligent behaviour and decision-making develop primarily from the interaction between brain, body and environment [8]. We use our entire environment, including integrated technologies, as an extended mind or memory storage [9], for example to facilitate knowledge retrieval and to reduce the cognitive demands of a task. Familiar technical aids include our fingers for counting, GPS devices for navigation, or the use of the internet for knowledge retrieval. We use such technical aids to either simplify or surrender tasks completely. Hence, we are permanently engaged in what can be called cognitive offloading [10].

16.3   Connecting Brain and Machine: Brain-Computer Interfaces

From a technological perspective, computers and machines are increasingly capable of learning, communicating and making decisions. Thus, the interaction between humans and technology gains additional dynamics. Speech or gesture and mimic recognition are increasingly replacing former input devices such as a mouse and keyboard. There is steady progress in the development of measurement techniques, sensor technologies and miniaturization of techniques for recording neurophysiological activity coupled with advanced signal processing, statistics and machine learning. Based on these developments, we gathered a tremendous understanding of cognitive functions and emotional processes underlying human behaviour, decision-making and social interactions over the last decades. Thus, brain and physiological signals allow us to derive many facets of subtle mental user states, like attention, affect, movement intention, cognitive workload and many more. One key invention for researching brain-based interactions between humans and machines is called Brain-Computer Interfaces (BCIs) [11, 12]. The BCI is currently the most direct form of an interface for the interaction between a user and a technical system.

16.4 Measurement Technologies and Applications of Brain-Computer Interfaces

The backbone of BCIs is technology for the real-time recording of neurophysiological activity that can be divided into invasive and non-invasive measurement techniques. *Invasive recording techniques* require brain surgery to implant electrodes directly into the brain. These recordings are further subdivided into brain-surface electrodes, like e.g. electrocorticography (ECoG) and brain-penetrating microelectrodes (for a comprehensive overview, see Thakor [13]). *Non-invasive recordings* for BCIs are subdivided into (a) portable measurement techniques, like electroencephalography (EEG) and functional near-infrared spectroscopy (fNIRS) and (b) stationary systems like magnetoencephalography (MEG) and functional magnetic resonance imaging (fMRI). EEG and MEG allow measuring the electromagnetic activity of multiple cortical neurons directly by recording voltage fluctuations via electrodes on the scalp (EEG) or by using very sensitive superconducting sensors (so-called SQUIDs, superconducting quantum interference device in MEG) [14, 15]. Both fMRI and fNIRS measure neuronal activity indirectly. These techniques record metabolic processes related to neuronal activity by capturing hemodynamic changes in the blood flow. Hence, they enable the precise localization of the activation and deactivation of certain brain regions [16, 17].

Since its beginnings in the 1970s [18], BCI research focussed primarily on clinical and medical applications. The main purpose was to provide users that have physical or perceptual limitations with a communication tool or to allow them to control a technical device, for example for locked-in or stroke patients. In such applications, certain mental states that the user voluntarily generated are decoded while circumventing any muscular activity [19–24]. BCIs enable the development of assistive and restorative technologies to control wheelchairs [25], orthoses [26–28], prostheses [29], service robots [30] and web-applications [31]. Besides active control for users with motor impairments, BCIs are also applicable for neurofeedback training, for example to treat patients with psychiatric disorders such as depression, schizophrenia or attentional deficits [32–36]. Advances in these fields have led to a boost in mobile technologies and sophisticated machine-learning algorithms that can be further exploited for monitoring healthy users and laying the basis for non-medical applications of BCIs [11, 12, 37, 38].

16.5   Neuroadaptive Technology and Its Potential for Future Human-Computer Interaction Applications

In our everyday life, technical systems are becoming more and more prominent and serve the purpose of supporting us in our daily routines. Interactive machines and adaptive systems obtain information from the user's interaction behaviour [39] through integrated sensors (e.g. smartphones) or environmental sensors (e.g. cameras) [40–42]. Such intelligent systems are summarized under the term context-aware systems [43, 44]. Context-aware systems are able to adapt the interaction based on the current context information (including information about the purpose of use, objective to be achieved, and tasks), thus making machines sensitive to physical environments, locations and situations. Examples range from very simple adaptations such as screen brightness to the current time of day and ambient lighting; lane-keeping and distance assistants in (autonomous) vehicles; and cooperating industrial robots and service robots for domestic use for the elderly.

However, the user with her individual preferences, skills and abilities receives less attention. In order to provide an optimal interaction between user and adaptive and autonomous technologies, it is important to take not only environmental and contextual conditions, but also the current user state (with her preferences and intentions) into account appropriately. Thus, machines need an understanding of the user, information about the user that goes beyond the bare necessities for controlling the machine. Therefore, it is a major prerequisite that technology reacts sensitively and promptly to its users, to create a collaborative and assistive interaction. Over the last years, the use of machine-learning algorithms for computational user modelling has increased substantially [45–47]. The basic idea is that computational user models represent more fine-graded aspects of the user, such as skills, preferences and cognitive abilities, as well as contextual changes such as selective attention, working memory load and the current emotional state and mood. It allows system adaptation to complex situations without inflexible dependence on predetermined programs [48, 49] and provides the basis for a symbiotic interaction between user and machine to collaborate and cooperate in making collective decisions.

For a long time, clinical and medical applications have been the primary goals of BCI research. In classical approaches, active control-based BCIs consider decoding brain activity to map it to commands that can drive an application that is running on a computer or a device. Examples for assistive or restorative BCIs are the P300 speller [50] that uses an event-related potential correlating with attentional resources, menu selection and exoskeleton control using Steady-State-Visual-Evoked-Potential (SSVEP) paradigms that are natural responses to visual stimuli at specific frequencies [51, 52], or binary selection through motor imagery paradigms that are produced by voluntarily modulating certain oscillatory sensorimotor rhythms [20, 26–28].

The introduction of passive BCIs [53] as a new concept coupled with new mobile and deployable sensor technologies for EEG and fNIRS [54–56] and advanced signal processing and machine-learning algorithms [37, 38, 57–60] for artefact correction and classification of cognitive and emotional states makes BCIs ready for non-clinical usage [11, 12, 61–63]. In classical BCIs, the machine-learning algorithms that are used focus mainly on the number of bits transmitted per minute and the successful classification rates of extracted brain patterns. In the passive BCI concept, the bit transmission rate is not the primary interest, but rather to focus on augmenting human-computer interaction. Hence, users do not need to carry out any mental actions actively to produce brain patterns that are translated to computer action. To the contrary, the passive BCI concept is envisioned as a continuous brain-monitoring process that is used to stratify the user according to his/her cognitive or emotional state. This provides the basis for extended computational user models in a human-machine control loop. Furthermore, this loop requires a precise knowledge of the psychological processes and corresponding neurophysiological correlates on which the adaptive computer system depends. Open-loop EEG or fNIRS-based passive BCIs to monitor psychological processes such as user engagement, user intention, selective attention, emotional engagement and workload in students, drivers, pilots or air traffic controllers have already been introduced [64–73]. Affective reactions such as valence and arousal are another source of user information that can serve as possible input to adaptive computer systems [61–63, 74, 75]. In a closed-loop human-computer interaction paradigm, this information enriches a user model to enable not only a concrete command, but also an adequate system adaptation to the user's preferences, skills and abilities. With the help of sophisticated signal processing and machine-learning techniques, neurophysiological signals can be interpreted in the sense of a continuous representation of the user's condition and provide information about psychological processes such as cognition, emotion and motivation. In a control loop, the estimated user model serves as an input variable in order to optimally support the goal of user interaction and certain user needs by intelligent system adaptation.

The developments in brain-based interactions enable the design of a neuroadaptive system loop [61–63, 76–78]. These loops are currently being discussed as a strategy to make adaptive and autonomous technologies more user-oriented and augment human-computer interaction. Possible future neuroadaptive applications, among others, are for example:

- Intelligent vehicles that dynamically adapt the level of automation of the driving task to the current intention, attentional or workload level of the driver [67, 79–82].
- Interactive e-learning programs that adapt speed and difficulty to the cognitive abilities of the user [73, 83].
- Neurofeedback-based interfaces to promote subjective well-being by training concentration and relaxation [84].
- Personalized internet applications that capture affective user reaction to adapt their content, presentation and interaction mechanisms to individual needs and preferences [61–63].

- Collaborative robots that react sensitively to user intentions, emotions and attentional levels [85–88].

There are two main benefits of implicit user interaction via a neuroadaptive control loop for personalized system applications. (1) *Complementarity*: There is no interference with other activities in the interaction cycle. The neuroadaptive control loop expands the communication between human and machine and thereby contributes to the self-learning process of the system towards individual user abilities, skills and preferences. (2) *Assistance for automation:* The neuroadaptive system can help to increase the situation-awareness of the user towards automated system behaviour. A transparent system behaviour develops by considering implicit user reactions during longer periods of interaction. Consequently, the user is expected to experience feelings of control over and trust in the automated system. Future research will reveal the extent to which these expectations are correct.

The increasing research interest in the still very young field of neuroadaptive technologies can also be observed at human factor engineering[1] and affective computing[2] conferences and newly emerging, popular conferences such as neuroergonomics[3] and neuroadaptive technology.[4] Innovation-friendly companies, e.g. from the automobile industry, already use brain and physiological methods for their consumer research. Furthermore, technology companies from Silicon Valley, like *Facebook*[5] or Elon Musk—founded *NeuraLink*[6]—invest in research on invasive closed-loop neuroadaptive technologies.

Although the use and advantages of neuroadaptive interaction are indisputable, there are still some major gaps and challenges to overcome before they can be applied outside of lab conditions. (1) *Understanding brain functions out of the lab:* There is still a significant lack of basic knowledge on human brain functions in complex real-world situations where individuals perform activities in natural environments and social contexts. In such environments, signal analysis and machine-learning interpretation is still very challenging. Robust algorithms to deal with real-world artefacts that strongly exceed the signals of interest are still needed. (2) *Integration of context information*: Neurophysiological signals cannot be interpreted in isolation, but must be analysed and classified in a given context of application. While context is controlled and known in the lab, real-life applications require the combination of context information to adapt technical systems to user states and social situations under real-world conditions.

---

[1] https://www.ahfe2019.org/, accessed 29th July 2019.

[2] http://acii-conf.org/2019/, accessed 29th July 2019.

[3] http://www.biomed.drexel.edu/neuroergonomics/, accessed 29th July 2019.

[4] http://neuroadaptive.org/conference, accessed 29th July 2019.

[5] https://www.scientificamerican.com/article/facebook-launches-moon-shot-effort-to-decode-speech-direct-from-the-brain/, accessed 29th July 2019.

[6] https://www.wsj.com/articles/elon-musk-launches-neuralink-to-connect-brains-with-computers-1490642652, accessed 29th July 2019.

Consequently, considerable research effort is needed to realize closed-loop solutions based on brain signals robust enough to deal with the high complexity of real-life applications [11, 12, 89].

The primary goal of a future neuroadaptive technology research agenda is to consistently align the increasing intelligence and autonomy of technical assistive systems with the emotional needs and cognitive abilities of the human—*a human-centred neuroadaptive technology research roadmap*. By their individual adaptability, neuroadaptive technologies contribute significantly to a human-centred, efficient and acceptable technology. The applied research in this field and the possible transfer into real-life applications requires a strong transdisciplinary research agenda. Due to its far-reaching social implications, research and developments does not only have to face technological, but also social challenges, like including questions about cognitive liberty, mental privacy, mental integrity and psychological integrity. In addition to computer science and neuroscience, the integration of further disciplines is needed, such as positive psychology that aims to foster human flourishing and well-being by researching positive emotions and its influences during human-technology interaction as well as ethics and social sciences. If neuroadaptive technologies are applied in non-medical areas, they must be consistently oriented to the needs and ethical values of the users and society.

References

1. Osiurak F, Navarro J, Reynaud E. How our cognition shapes and is shaped by technology: a common framework for understanding human tool-use interactions in the past, present, and future. Front Psychol. 2018;9:293. https://doi.org/10.3389/fpsyg.2018.00293.
2. Moore P. Enhancing me: the hope and the hype of human enhancement (TechKnow). New York: Wiley; 2008.
3. Kahneman D. Objective happiness. New York: Russell Sage Foundation; 1999. p. xii. 593 pp.
4. Jameson A. Understanding and dealing with usability side effects of intelligent processing. AI Mag. 2009;30:23–40.
5. Deci EL, Ryan RM, editors. Handbook of self-determination research. Softcover ed. Rochester: University of Rochester Press; 2004.
6. Hassenzahl M. User experience (UX): towards an experiential perspective on product quality. New York: ACM Press; 2008. p. 11.
7. Spath D, Peissner M, Sproll S. Methods from neuroscience for measuring user experience in work environments. In: Rice V, editor. Advances in understanding human performance. Boca Raton: CRC Press; 2010. p. 111–21.
8. Engel AK, Maye A, Kurthen M, König P. Where's the action? The pragmatic turn in cognitive science. Trends Cogn Sci. 2013;17:202–9.
9. Wilson M. Six views of embodied cognition. Psychon Bull Rev. 2002;9:625–36.
10. Risko EF, Gilbert SJ. Cognitive offloading. Trends Cogn Sci. 2016;20:676–88.
11. Blankertz B, Acqualagna L, Dähne S, Haufe S, Schultze-Kraft M, Sturm I, Ušćumlic M, Wenzel MA, Curio G, Müller K-R. The Berlin brain-computer interface: progress beyond communication and control. Front Neurosci. 2016;10:530. https://doi.org/10.3389/fnins.2016.00530.
12. Cinel C, Valeriani D, Poli R. Neurotechnologies for human cognitive augmentation: current state of the art and future prospects. Front Hum Neurosci. 2019;13:31. https://doi.org/10.3389/fnhum.2019.00013.
13. Thakor NV. Translating the brain-machine interface. Sci Transl Med. 2013;5:210ps17.

14. Nunez PL, Srinivasan R. Electric fields of the brain: the neurophysics of EEG. 2nd ed. Oxford: Oxford University Press; 2006.
15. Hämäläinen M, Hari R, Ilmoniemi RJ, Knuutila J, Lounasmaa OV. Magnetoencephalography—theory, instrumentation, and applications to noninvasive studies of the working human brain. Rev Mod Phys. 1993;65:413–97.
16. Logothetis NK, Pauls J, Augath M, Trinath T, Oeltermann A. Neurophysiological investigation of the basis of the fMRI signal. Nature. 2001;412:150–7.
17. Ferrari M, Quaresima V. A brief review on the history of human functional near-infrared spectroscopy (fNIRS) development and fields of application. NeuroImage. 2012;63:921–35.
18. Vidal JJ. Toward direct brain-computer communication. Annu Rev Biophys Bioeng. 1973;2:157–80.
19. Birbaumer N, Ghanayim N, Hinterberger T, Iversen I, Kotchoubey B, Kübler A, Perelmouter J, Taub E, Flor H. A spelling device for the paralysed. Nature. 1999;398:297–8.
20. Ramos-Murguialday A, Broetz D, Rea M, et al. Brain-machine interface in chronic stroke rehabilitation: a controlled study: BMI in chronic stroke. Ann Neurol. 2013;74:100–8.
21. Kübler A, Nijboer F, Mellinger J, Vaughan TM, Pawelzik H, Schalk G, McFarland DJ, Birbaumer N, Wolpaw JR. Patients with ALS can use sensorimotor rhythms to operate a brain-computer interface. Neurology. 2005;64:1775–7.
22. Münßinger JI, Halder S, Kleih SC, Furdea A, Raco V, Hösle A, Kübler A. Brain painting: first evaluation of a new brain–computer interface application with ALS-patients and healthy volunteers. Front Neurosci. 2010;4:182. https://doi.org/10.3389/fnins.2010.00182.
23. Wolpaw JR, Birbaumer N, McFarland DJ, Pfurtscheller G, Vaughan TM. Brain-computer interfaces for communication and control. Clin Neurophysiol. 2002;113:767–91.
24. Wolpaw JR. Brain-computer interfaces as new brain output pathways. J Physiol Lond. 2007;579:613–9.
25. Carlson T, JdR M. Brain-controlled wheelchairs: a robotic architecture. IEEE Robot Autom Mag. 2013;20:65–73.
26. Vukelić M, Gharabaghi A. Oscillatory entrainment of the motor cortical network during motor imagery is modulated by the feedback modality. NeuroImage. 2015;111:1–11.
27. Brauchle D, Vukelić M, Bauer R, Gharabaghi A. Brain state-dependent robotic reaching movement with a multi-joint arm exoskeleton: combining brain-machine interfacing and robotic rehabilitation. Front Hum Neurosci. 2015;9:564. https://doi.org/10.3389/fnhum.2015.00564.
28. Vukelić M, Belardinelli P, Guggenberger R, Royter V, Gharabaghi A. Different oscillatory entrainment of cortical networks during motor imagery and neurofeedback in right and left handers. NeuroImage. 2019;195:190–202.
29. Rohm M, Schneiders M, Müller C, Kreilinger A, Kaiser V, Müller-Putz GR, Rupp R. Hybrid brain–computer interfaces and hybrid neuroprostheses for restoration of upper limb functions in individuals with high-level spinal cord injury. Artif Intell Med. 2013;59:133–42.
30. Leeb R, Tonin L, Rohm M, Desideri L, Carlson T, JdR M. Towards independence: a BCI telepresence robot for people with severe motor disabilities. Proc IEEE. 2015;103:969–82.
31. Bensch M, Karim AA, Mellinger J, Hinterberger T, Tangermann M, Bogdan M, Rosenstiel W, Birbaumer N. Nessi: an EEG-controlled web browser for severely paralyzed patients. Comput Intell Neurosci. 2007;2007:1–5.
32. Wyckoff S, Birbaumer N. Neurofeedback and brain-computer interfaces. In: Mostofsky DI, editor. The handbook of behavioral medicine. Oxford: Wiley; 2014. p. 275–312.
33. Birbaumer N, Ruiz S, Sitaram R. Learned regulation of brain metabolism. Trends Cogn Sci (Regul Ed). 2013;17:295–302.
34. Ruiz S, Lee S, Soekadar SR, Caria A, Veit R, Kircher T, Birbaumer N, Sitaram R. Acquired self-control of insula cortex modulates emotion recognition and brain network connectivity in schizophrenia. Hum Brain Mapp. 2013;34:200–12.
35. Choi SW, Chi SE, Chung SY, Kim JW, Ahn CY, Kim HT. Is alpha wave neurofeedback effective with randomized clinical trials in depression? A pilot study. Neuropsychobiology. 2011;63:43–51.
36. Ehlis A-C, Schneider S, Dresler T, Fallgatter AJ. Application of functional near-infrared spectroscopy in psychiatry. NeuroImage. 2014;85:478–88.

37. Craik A, He Y, Contreras-Vidal JL. Deep learning for electroencephalogram (EEG) classification tasks: a review. J Neural Eng. 2019;16:031001.
38. Lotte F, Bougrain L, Cichocki A, Clerc M, Congedo M, Rakotomamonjy A, Yger F. A review of classification algorithms for EEG-based brain–computer interfaces: a 10 year update. J Neural Eng. 2018;15:031005.
39. Seifert C, Granitzer M, Bailer W, Orgel T, Gantner L, Kern R, Ziak H, Petit A, Schlötterer J, Zwicklbauer S. Ubiquitous access to digital cultural heritage. J Comput Cult Herit. 2017;10:1–27.
40. Radu V, Lane ND, Bhattacharya S, Mascolo C, Marina MK, Kawsar F. Towards multimodal deep learning for activity recognition on mobile devices. In: Proceedings of the 2016 ACM international joint conference on pervasive and ubiquitous computing adjunct—UbiComp'16. Heidelberg: ACM Press; 2016. p. 185–8.
41. Sankaran K, Zhu M, Guo XF, Ananda AL, Chan MC, Peh L-S. Using mobile phone barometer for low-power transportation context detection. In: Proceedings of the 12th ACM conference on embedded network sensor systems—SenSys'14. Memphis: ACM Press; 2014. p. 191–205.
42. Liu H, Wang J, Wang X, Qian Y. iSee: obstacle detection and feedback system for the blind. In: Proceedings of the 2015 ACM international joint conference on pervasive and ubiquitous computing and proceedings of the 2015 ACM international symposium on wearable computers—UbiComp'15. Osaka: ACM Press; 2015. p. 197–200.
43. Mens K, Capilla R, Cardozo N, Dumas B. A taxonomy of context-aware software variability approaches. In: Companion proceedings of the 15th international conference on modularity—MODULARITY companion 2016. Malaga: ACM Press, Spain; 2016. p. 119–24.
44. Kaklanis N, Biswas P, Mohamad Y, Gonzalez MF, Peissner M, Langdon P, Tzovaras D, Jung C. Towards standardisation of user models for simulation and adaptation purposes. Univ Access Inf Soc. 2016;15:21–48.
45. Yan L, Ma Q, Yoshikawa M. Classifying twitter users based on user profile and followers distribution. In: Decker H, Lhotská L, Link S, Basl J, Tjoa AM, editors. Database and expert systems applications. Berlin: Springer; 2013. p. 396–403.
46. Gao R, Hao B, Bai S, Li L, Li A, Zhu T. Improving user profile with personality traits predicted from social media content. In: Proceedings of the 7th ACM conference on recommender systems—RecSys'13. Hong Kong: ACM Press; 2013. p. 355–8.
47. Besel C, Schlötterer J, Granitzer M. On the quality of semantic interest profiles for onine social network consumers. SIGAPP Appl Comput Rev. 2016;16:5–14.
48. Licklider JCR. Man-computer Symbiosis. IRE Trans Hum Factors Electron HFE. 1960;1:4–11.
49. Pope AT, Bogart EH, Bartolome DS. Biocybernetic system evaluates indices of operator engagement in automated task. Biol Psychol. 1995;40:187–95.
50. Krusienski DJ, Sellers EW, McFarland DJ, Vaughan TM, Wolpaw JR. Toward enhanced P300 speller performance. J Neurosci Methods. 2008;167:15–21.
51. Kwak N-S, Müller K-R, Lee S-W. A lower limb exoskeleton control system based on steady state visual evoked potentials. J Neural Eng. 2015;12:056009.
52. Yin E, Zhou Z, Jiang J, Chen F, Liu Y, Hu D. A novel hybrid BCI speller based on the incorporation of SSVEP into the P300 paradigm. J Neural Eng. 2013;10:026012.
53. Zander TO, Kothe C. Towards passive brain-computer interfaces: applying brain-computer interface technology to human-machine systems in general. J Neural Eng. 2011;8:025005.
54. McDowell K, Lin C-T, Oie KS, Jung T-P, Gordon S, Whitaker KW, Li S-Y, Lu S-W, Hairston WD. Real-world neuroimaging technologies. IEEE Access. 2013;1:131–49.
55. Zander TO, Andreessen LM, Berg A, Bleuel M, Pawlitzki J, Zawallich L, Krol LR, Gramann K. Evaluation of a dry EEG system for application of passive brain-computer interfaces in autonomous driving. Front Hum Neurosci. 2017;11:78. https://doi.org/10.3389/fnhum.2017.00078.
56. Piper SK, Krueger A, Koch SP, Mehnert J, Habermehl C, Steinbrink J, Obrig H, Schmitz CH. A wearable multi-channel fNIRS system for brain imaging in freely moving subjects. NeuroImage. 2014;85:64–71.
57. Haeussinger FB, Dresler T, Heinzel S, Schecklmann M, Fallgatter AJ, Ehlis A-C. Reconstructing functional near-infrared spectroscopy (fNIRS) signals impaired by extra-cranial confounds: an easy-to-use filter method. NeuroImage. 2014;95:69–79.

58. Schecklmann M, Mann A, Langguth B, Ehlis A-C, Fallgatter AJ, Haeussinger FB. The temporal muscle of the head can cause artifacts in optical imaging studies with functional near-infrared spectroscopy. Front Hum Neurosci. 2017;11:456. https://doi.org/10.3389/fnhum.2017.00456.
59. Biessmann F, Plis S, Meinecke FC, Eichele T, Müller K-R. Analysis of multimodal neuroimaging data. IEEE Rev Biomed Eng. 2011;4:26–58.
60. Dahne S, BieBmann F, Meinecke FC, Mehnert J, Fazli S, Mtuller K-R. Multimodal integration of electrophysiological and hemodynamic signals. IEEE; 2014. p. 1–4.
61. Bauer W, Vukelić M. EMOIO research project: an interface to the world of computers. In: Neugebauer R, editor. Digital transformation. Berlin: Springer; 2019. p. 129–44.
62. Vukelić M, Pollmann K, Peissner M. Toward brain-based interaction between humans and technology. In: Neuroergonomics. Amsterdam: Elsevier; 2019. p. 105–9.
63. Pollmann K, Ziegler D, Peissner M, Vukelić M. A new experimental paradigm for affective research in neuro-adaptive technologies. New York: ACM Press; 2017. https://doi.org/10.1145/3038439.3038442.
64. Dijksterhuis C, de Waard D, Brookhuis KA, Mulder BLJM, de Jong R. Classifying visuomotor workload in a driving simulator using subject specific spatial brain patterns. Front Neurosci. 2013;7:149. https://doi.org/10.3389/fnins.2013.00149.
65. Berka C, Levendowski DJ, Lumicao MN, Yau A, Davis G, Zivkovic VT, Olmstead RE, Tremoulet PD, Craven PL. EEG correlates of task engagement and mental workload in vigilance, learning, and memory tasks. Aviat Space Environ Med. 2007;78:B231–44.
66. Aricò P, Borghini G, Di Flumeri G, Colosimo A, Pozzi S, Babiloni F. A passive brain–computer interface application for the mental workload assessment on professional air traffic controllers during realistic air traffic control tasks. In: Progress in brain research. Amsterdam: Elsevier; 2016. p. 295–328.
67. Haufe S, Kim J-W, Kim I-H, Sonnleitner A, Schrauf M, Curio G, Blankertz B. Electrophysiology-based detection of emergency braking intention in real-world driving. J Neural Eng. 2014;11:056011.
68. Lahmer M, Glatz C, Seibold VC, Chuang LL. Looming auditory collision warnings for semiautomated driving: an ERP Study. In: Proceedings of the 10th international conference on automotive user interfaces and interactive vehicular applications—automotiveUI'18. Toronto: ACM Press. 2018. p. 310–9.
69. Ihme K, Unni A, Zhang M, Rieger JW, Jipp M. Recognizing frustration of drivers from face video recordings and brain activation measurements with functional near-infrared spectroscopy. Front Hum Neurosci. 2018;12:327.
70. Dehais F, Roy RN, Scannella S. Inattentional deafness to auditory alarms: inter-individual differences, electrophysiological signature and single trial classification. Behav Brain Res. 2019;360:51–9.
71. Dehais F, Duprès A, Blum S, Drougard N, Scannella S, Roy R, Lotte F. Monitoring Pilot's mental workload using ERPs and spectral power with a six-dry-electrode EEG system in real flight conditions. Sensors. 2019;19:1324.
72. Ayaz H, Shewokis PA, Bunce S, Izzetoglu K, Willems B, Onaral B. Optical brain monitoring for operator training and mental workload assessment. NeuroImage. 2012;59:36–47.
73. Walter C, Rosenstiel W, Bogdan M, Gerjets P, Spüler M. Online EEG-based workload adaptation of an arithmetic learning environment. Front Hum Neurosci. 2017;11:286.
74. Mühl C, Allison B, Nijholt A, Chanel G. A survey of affective brain computer interfaces: principles, state-of-the-art, and challenges. Brain Comput Interfaces. 2014;1:66–84.
75. Liberati G, Federici S, Pasqualotto E. Extracting neurophysiological signals reflecting users' emotional and affective responses to BCI use: a systematic literature review. NeuroRehabilitation. 2015;37:341–58.
76. Zander TO, Krol LR, Birbaumer NP, Gramann K. Neuroadaptive technology enables implicit cursor control based on medial prefrontal cortex activity. Proc Natl Acad Sci. 2016;113(52):14898–903.
77. Fairclough SH. Fundamentals of physiological computing. Interact Comput. 2009;21:133–45.
78. Hettinger LJ, Branco P, Encarnacao LM, Bonato P. Neuroadaptive technologies: applying neuroergonomics to the design of advanced interfaces. Theor Issues Ergon Sci. 2003;4:220–37.

79. Sonnleitner A, Simon M, Kincses WE, Buchner A, Schrauf M. Alpha spindles as neurophysi-ological correlates indicating attentional shift in a simulated driving task. Int J Psychophysiol. 2012;83:110–8.
80. Ricardo Chavarriaga LG. Detecting cognitive states for enhancing driving experience. In: International BCI meeting brain-computer interface 2013 proceedings of the fifth international brain-computer Interface meeting: defining the future June 3-7 2013 Asilomar conference cen-ter, Pacific grove, California, USA; 2015. https://doi.org/10.3217/978-3-85125-260-6-60.
81. Unni A, Ihme K, Jipp M, Rieger JW. Assessing the driver's current level of working memory load with high density functional near-infrared spectroscopy: a realistic driving simulator study. Front Hum Neurosci. 2017;11:167.
82. Pollmann K, Stefani O, Bengsch A, Peissner M, Vukelić M. How to work in the car of the future?: a neuroergonomical study assessing concentration, performance and workload based on subjective, behavioral and neurophysiological insights. In: Proceedings of the 2019 CHI conference on human factors in computing systems—CHI'19. Glasgow: ACM Press; 2019. p. 1–14.
83. Spüler M, Krumpe T, Walter C, Scharinger C, Rosenstiel W, Gerjets P. Brain-computer inter-faces for educational applications. In: Buder J, Hesse FW, editors. Informational environ-ments. Cham: Springer International Publishing; 2017. p. 177–201.
84. Kosuru RK, Lingelbach K, Bui M, Vukelić M. MindTrain: how to train your mind with inter-active technologies. In: Proceedings of mensch und computer 2019 on—MuC'19. Hamburg: ACM Press; 2019. p. 643–7.
85. Perrin X, Chavarriaga R, Colas F, Siegwart R, Millán JR. Brain-coupled interaction for semi-autonomous navigation of an assistive robot. Roboti Auton Syst. 2010;58:1246–55.
86. Chavarriaga R, Sobolewski A, Millãjn JdR. Errare machinale est: the use of error-related potentials in brain-machine interfaces. Front Neurosci. 2014;8:208. https://doi.org/10.3389/fnins.2014.00208.
87. Iwane F, Halvagal MS, Iturrate I, Batzianoulis I, Chavarriaga R, Billard A, Millan JdR. Inferring subjective preferences on robot trajectories using EEG signals. In: 2019 9th international IEEE/EMBS conference on neural engineering (NER). San Francisco: IEEE; 2019. p. 255–8.
88. Edelman BJ, Meng J, Suma D, Zurn C, Nagarajan E, Baxter BS, Cline CC, He B. Noninvasive neuroimaging enhances continuous neural tracking for robotic device control. Sci Robot. 2019;4:eaaw6844.
89. Brouwer A-M, Zander TO, van Erp JBF, Korteling JE, Bronkhorst AW. Using neurophysi-ological signals that reflect cognitive or affective state: six recommendations to avoid common pitfalls. Front Neurosci. 2015;9:136. https://doi.org/10.3389/fnins.2015.00136.

Printed in the United States
by Baker & Taylor Publisher Services